Leandro Danielski

Supercritical Processing of Plant Materials

Leandro Danielski

Supercritical Processing of Plant Materials

Applications to the Extraction and Fractionation of Natural Products

VDM Verlag Dr. Müller

Imprint

Bibliographic information by the German National Library: The German National Library lists this publication at the German National Bibliography; detailed bibliographic information is available on the Internet at http://dnb.d-nb.de.
Any brand names and product names mentioned in this book are subject to trademark, brand or patent protection and are trademarks or registered trademarks of their respective holders. The use of brand names, product names, common names, trade names, product descriptions etc. even without a particular marking in this works is in no way to be construed to mean that such names may be regarded as unrestricted in respect of trademark and brand protection legislation and could thus be used by anyone.

Cover image: www.purestockx.com

Publisher:
VDM Verlag Dr. Müller Aktiengesellschaft & Co. KG, Dudweiler Landstr. 125 a, 66123 Saarbrücken, Germany,
Phone +49 681 9100-698, Fax +49 681 9100-988,
Email: info@vdm-verlag.de

Produced in USA and UK by:
Lightning Source Inc., La Vergne, Tennessee, USA
Lightning Source UK Ltd., Milton Keynes, UK
BookSurge LLC, 5341 Dorchester Road, Suite 16, North Charleston, SC 29418, USA

ISBN: 978-3-639-02458-6

Acknowledgements

This work presents the results obtained during my period of four years (from April 2003 to March 2007) at the Institut für Thermische Verfahrenstechnik of the Technische Universität Hamburg-Harburg.

First of all, I would like to thank Prof. Dr.-Ing. Gerd Brunner for accepting and incorporating me into his research program. Additionally, I am grateful for his guidance, attention, patience, support, encouragement and for the proposal of an interesting research thema. All possible conditions necessary for a good work were provided without restraints.

I thank Prof. Dr.-Ing. G. Fieg and Prof. Dr.rer.nat. A. Liese for taking part on the evaluation of this work and for additional reporting.

The engagement of some undergraduate students provided interesting discussions and contributions. I am glad to have worked with T. C. Bandiera, L. F. Picolo, C. Schwänke, D. Tjahjasari and K. A. Tsankov. Thank you all.

I would like to thank all co-workers and technicians at the Institute for the good treatment dispensed to myself and for a gentle work atmosphere, especially during my initial times in Hamburg. During my first months at the Institute, the help of D. Luz da Silva and C. Zetzl must be acknowledged. I am grateful to all my office colleagues, A. Bezold, K. Gast, K. Rosenkranz, R. Schreiber and W. S. Long. Thank you for your contributions and good humor. Special mention must be made to M. H. Chuang, S. Meyer-Storckmann and A. Paiva for their support in several aspects. The technical support of R. Henneberg, M. Kammlott, F. Sokolinski and T. Weselmann is also acknowledged. The contact with several exchange students and especial guests must also be mentioned. I am glad to have met such interesting and competent people, like M. Aungsukiatethavorn, R. M. Barros, P. R. Calvo, P. Chan, S. Fajar, L. Filipova, K. Gerasimov, S. Lohner, C. Möbius, V. Panayiotou, V. Patil, S. Rodriguez-Rojo, R. Ruivo, A. Serbanovic, P. Toshew, E. Uquiche and E. Vaquero.

The incentive and contributions of my former Brazilian professors from Florianópolis, Prof. Dr. S. R. S. Ferreira and Dr.-Ing. H. Hense, must not be forgotten. My gratitude and respect for them are unquestionable. In addition, I would like to thank my former Master colleagues, L. M. A. S. Campos, A. K. Genena and E. M. Z. Michielin.

My gratitude must be also extended to my Brazilian friends in Hamburg and across Germany. During my German language course in Göttingen I met most of them. Thanks to my dear Brazilian NIT family, B. Almeida, A. R. Barreto, F. Bertocco, G. Cenachi, M. Dytz,

O. M. Ferri, M. Ghislandi, G. M. Gualberto, S. Kuester, J. C. Longo, R. Martins, F. A. Neto and G. Wiggers. Thank you L. K. Luna, L. P. Soares, K. Trefflich and little Luan for the time we spent together and for your company.

My gratitude is also expressed to E. Wiegmann and L. Wiegmann, for treating me as a member of their family.

Nothing of this was possible without the help and encouragement provided by my family and friends in Brazil. I am undoubtly grateful to my lovely parents and brothers.

Last but not least, the financial support of CAPES (Coordenação de Aperfeiçoamento de Pessoal de Nível Superior, Brasília, Brazil) under grant BEX 0816/02-7 during my period in Hamburg is gratefully acknowledged. I would like also to thank DAAD (Deutscher Akademischer Austauschdienst, Bonn, Germany) for providing me the chance to have the first contacts with the German language and culture at Goethe Institut (Göttingen).

Leandro Danielski – Hamburg, 02.07.2007

To my family and to my dear
friend Gisielly Schoeffel (*in memoriam*).

Contents

Symbols and Abbreviations

Latin Symbols

A	-	Peak area
A	m^2	Cross sectional area
a	m^{-1}	Specific surface of solid phase
b	-	Constant of the Langmuir isotherm and of the Logistic model
Bi	-	Biot number
C, c	wt.-% or g/g	Concentrations
D	m^2/s	Diffusion coefficient
d	m	Diameter
E	%	Degree of Extraction
F	-	Feed
f	Pa	Fugacity of a determined component
J	kg/m^2	Mass transfer rate
K	-	Distribution coefficient of one component
k_{ads}	-	Adsorption coefficient
k_{des}	-	Desorption coefficient
kf	-	Mass transfer coefficient
L, z	m	Column length/height, axial coordinate
L	-	Liquid phase, liquid flow
m	kg or g	Mass
n_{th}	-	Number of theoretical separation stages
P	MPa	Pressure
Pe	-	Peclet number
P_E	-	Load in Jänecke diagram (vapor phase)
P_R	-	Load in Jänecke diagram (liquid phase)
p, n	-	Stages in the balance of a countercurrent column
Q	-	Loading at full coverage (Langmuir isotherm)
q_e	-	Loading at equilibrium (Langmuir isotherm)
r	m	Radius
R	-	Reflux or raffinate flows
Re	-	Reynolds number
S	-	Solvent
Sc	-	Schmidt number
Sh	-	Sherwood number
t	s	Time
T	° C or K	Temperature
u	m/s	Interstitial velocity
V	l	Volume
V	-	Vapor phase, vapor flow
x	-	Fraction of a component in the liquid phase
y	-	Fraction of a component in the vapor (gas) phase

Greek letters

β	m²/s , -	Mass transport coefficient, selectivity
ε	-	Porosity, void fraction
η	g/(cm.s)	Dynamic viscosity
μ	-	Chemical potential
ν	-	Reflux ratio
ρ	kg/m³	Density
π	-	Number of phases coexisting in one system
δ	-	Partial derivative
α	-	Separation factor
Δ	-	Variation of one specific parameter or property

Indices

A	Component A; Aroma fraction
ax	Axial
B	Component B
C	Critical value
E, Ext	Extract
e, eq	Equilibrium condition
eff	Effective
F	Feed
f	Fluid
FFA	Free fatty acids
i	Component i
j	Component j
lim	Limonene
lin	Linalool
liq	Liquid
OR	Oryzanol(s)
p	Particle
p, n	Stages in the countercurrent column balance
R, Raf	Raffinate
sat	Saturation condition
ST	Sterols
T	Terpene(s)
t	Time
TG	Triglycerides
tpl	Terpinolene
0	Initial condition
1	Component 1
2	Component 2

Abbreviations

CP	Critical point
DD	Deodorizer distillates
Ex	Extract
FAME	Fatty acid methyl esters
FAO	Food and Agriculture Organization of the United Nations
FDA	Food and Drug Administration (USA)
FFA	Free fatty acids
FOS	Food Oil Sensor
FR	Folding ratio
GC	Gas Chromatography
GC-MS	Gas Chromatography-Mass Spectrometry
GRAS	"Generally Recognized As Safe"
HAP	Hazardous Air Pollutants
HETS	Height equivalent to one theoretical stage
HPLC	High Performance Liquid Chromatography
HSBO	Hydrogenated soybean oil
HVC	High volatile components
ISTD	Internal standards
LC	Liquid Chromatography
LDL-C	Low density lipoprotein cholesterol
LM	Logistic model
LVC	Low volatile components
MNMA	Methyl-N-methyl-anthranilate
MPO	Mandarin peel oil
OEC	Overall extraction curve(s)
OR	Oryzanol(s)
PSA	Pressure Swing Adsorption
PUFA	Polyunsaturated fatty acids
Raf	Raffinate
RBO	Rice bran oil
RF	Response Factor
RR	Reflux Ratio
RT	Retention Time
$SC\text{-}CO_2$	Supercritical carbon dioxide
SCF	Supercritical fluid
SFE	Supercritical Fluid Extraction
SFR	Solvent-to-feed ratio
ST	Sterols
TG	Triglycerides
UV-(A/B)	Ultraviolet radiation
VLE	Vapor-liquid equilibria
wt.-%	Weight percent

Summary

The aim of the present work was the measurement and modeling of multicomponent mixtures under supercritical conditions. The extraction and fractionation of natural organic components from two different plant materials using carbon dioxide as solvent at high pressures were investigated. These materials are usually treated as waste, being used as animal feed and for composting purposes.

The first material investigated was mandarin peel oil (MPO). Producers of essential peel oils (especially in Latin America) are involved with high energy demands and environmental aspects related to the conventional extraction and purification methods, what can be responsible for the thermal degradation of important substances and the contamination with residual solvents. Good results could provide them a guide for the application of an alternative technology in order to produce high value-added products for perfume and flavor industries starting from a by-product of fruits processing.

Two crude MPO samples were purchased directly from the producers: a red oil from Spain and a green oil from Brazil. Countercurrent experiments with high-pressurized CO_2 were carried out at pressures varying from 8 to 11.5 MPa and at 50, 60 and 70 °C, in order to separate the undesired terpene fraction (mostly limonene, approx. 95 wt.-%) from the raffinate one, which was collected at the bottom of the column and contained the active principles of the oil (aromatic substances). Applying countercurrent gas extraction, high selectivities between terpene and aroma fraction were obtained. The selectivities depended on the concentration and composition of the aroma fraction. Adsorption (using silica gel as adsorbent) followed by a selective desorption at different pressure levels produced two fractions of high purity. A temperature of 40 °C and pressure levels of 8 and 20 MPa for the desorption of terpenes and aroma components, respectively, were determined as optimal conditions. Scale-up experiments were also performed in order to achieve a higher enrichment of the desired aromatic substances, proving that supercritical extraction techniques can be perfectly employed for the deterpenation of citrus oils and selective fractionation of aromatic components.

In order to continue the investigation on the use of agricultural wastes as a source of nutritional compounds and food suplements, the second plant material investigated was rice bran oil (RBO). RBO is obtained from the rice bran, a by-product obtained through the polishing of rice grains. The bran corresponds to approx. 8 wt.-% of the grain and contains approx. 15-20 wt.-% oil. Nowadays, the world rice bran production reaches approx. 40 million metric tons/year and the high potential for extracting high value-added products from rice bran – such as phytosterols, triglycerides (TG), free fatty acids (FFA) and oryzanols – is already known. Therefore, the effect of experimental conditions on the high-pressurized CO_2 ability to extract and refine crude RBO was investigated.

RBO extraction and fractionation were carried out at different operational conditions, from 10 to 40 MPa and from 40 to 60 °C. Batch extractions have been modeled considering the extraction of more than one compound. In fact, 4 classes of components have been chosen: FFA, TG, sterols and oryzanols. The goal was to develop simulations of the multicomponent behavior in order to validate the experimental results.

Phase equilibrium measurements of crude RBO were also investigated. In order to perform the optimization of the operational conditions, the results obtained were used as a basis for the separation analysis, especially between FFA and TG. The chosen experimental parameters were pressure and temperature, reflux and solvent-to-feed ratios, and the number of separation stages. Based on these experimental data, the design of a production scale countercurrent column was possible on the basis of the results of a pilot scale column with an effective separation height of 6 m. The refining of crude RBO was successfully achieved, the extract fractions were enriched up to approx. 95 wt.-% FFA and the concentration in the raffinate fractions was below 1 wt.-%, confirming that SFE can be chosen as an alternative technique for the refining of vegetable oils.

1. Introduction

In the last decades, supercritical fluid extraction (SFE) has received especial attention in the field of solid material extractions and the fractionation of liquid mixtures. Nowadays, the possibility of extracting and fractionating vegetable oils receives widespread interest due to the direct applications in the food and pharmaceutical industries of the high-value products generated. Due to environmental and food processing regulations, supercritical fluids are a very attractive alternative to the conventional extraction and refining methods, such as extraction with organic solvents and vacuum and steam distillations. These conventional methods are usually carried out at high temperatures, what can be responsible for the destruction of valuable substances. Additionally, the use of organic solvents can also lead to product contamination with solvent residues.

Agricultural processes result in several by-products, which ones can present significant potential value. By-products are often under utilized (used as animal feed, for instance), and therefore their high nutritional value is lost. The problem faced by several industries consists of their inability to re-use these by-products. The feasibility of using food processing by-products and wastes for the recovery of high-value substances is affected by a number of factors, including physical and chemical characteristics, seasonality and quantity produced. Natural components can be found in food processing by-products and, with additional processing, can change from a low-value status to a high revenue stream of nutraceutical materials. Examples of these components include phytosterols, TG, vitamins and aromatic components, which can be used in the food industries as food supplements and/or food additives.

In this work, two different agricultural by-products were used as raw material. The possibilities of employing supercritical carbon dioxide (SC-CO_2) for the fractionation of mandarin peel oil (MPO) and for the extraction and fractionation of rice bran oil (RBO) were investigated.

Crude MPO was the first raw material investigated. The separation of terpenes from oxygenated aroma components was performed through countercurrent and ad-/desorption experiments. Both methods yielded different concentrated fractions, composed by the undesired terpene fraction and the highly desired aroma components. Additionally, adsorption/desorption scale-up experiments were performed.

The second agricultural by-product investigated was RBO, being extracted at different experimental conditions. This process has been modeled for more than one component, using models presented in the literature. Due to the high amount of undesired FFA present in crude RBO (from 7 to 10 wt.-%), countercurrent extraction was employed in order to separate the FFA fraction from low volatile components, composed mainly by TG, sterols and oryzanols. Based on phase equilibrium data, a separation analysis for this separation task could be performed and the results were verified through experiments using pilot-scale countercurrent columns.

1.1. Structure of the Work

This work is structured in 8 Chapters. A theoretical background related to the supercritical state and supercritical fluids, characteristics of gas extraction processes, as well as an overview on the applications of supercritical fluids are presented in Chapter 2. Chapter 2 also covers principles of phase equilibria, modeling multistage countercurrent processes, concepts of ad- and desorption and the mathematical modeling of SFE processes. Information about both raw materials used in this work (RBO and MPO) are summarized in Chapters 3 and 4, including a description of the chemical components evaluated, conventional extraction and refining methods, and a literature review on SFE applications. Analytical methods and experimental set-ups are described in Chapter 5. The results obtained for the deterpenation of MPO by employing countercurrent extraction and ad-/desorption are presented in Chapter 6. Chapter 7 summarizes the results obtained for RBO, covering phase equilibria measurements, modeling of multicomponent behavior and countercurrent experiments. Finally, conclusions and outlook of this work are presented in Chapter 8.

2. Theoretical Background

2.1. The Supercritical State

A substance reaches the so-called supercritical state when pressure and temperature are beyond their respective critical values. The critical point is defined as the end point of the vapor-pressure curve of the fluid, as presented in Figure 2.1. No distinction between liquid and vapor phase can be observed when a substance reaches its critical point and also overcomes it. At the critical point, it can be observed that the interface between vapor and liquid phases disappears. Supercritical fluids can vary easily from high to low density states without phase transition. Indeed, their thermodynamic and transport properties can be changed drastically only by manipulating the operational pressure and temperature conditions [1, 2]. The transition from gas-phase boundary to the supercritical phase is smooth and, because of that, models and basic considerations applied originally to liquid-vapor and liquid-liquid separations, such as solvent extraction, adsorption, desorption, rectification, stripping and distillation, may also be applicable to supercritical fluid extraction processes [1].

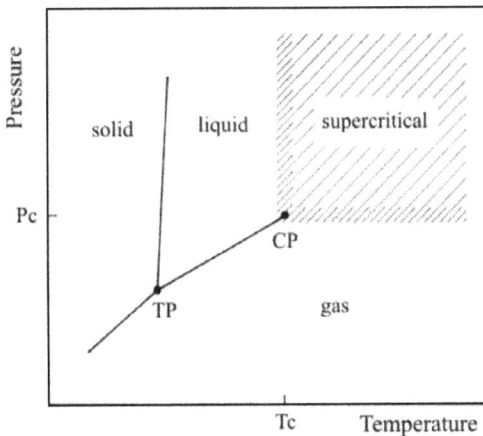

Figure 2.1. P-T diagram of a pure substance (adapted from [1]).

As can be seen from Table 2.1, supercritical fluids (SCF) present liquid-like densities, gas-like viscosities and diffusion coefficients located in the range between gas and liquid states. Due to these unique and advantageous features, namely high solvent power associated with gas-like transport properties, supercritical fluids have been chosen for several engineering applications [1, 2], what will be presented in Chapter 2.1.2.

Table 2.1. Physical properties of supercritical fluids [1].

Fluid	P/ T (MPa)/(K)	Density ρ (kg/m^3)	Diffusion coefficient D_{ij} (cm^2/s)	Viscosity η (g/cm·s)
Gas	0.1/298	0.6-2.0	0.1-0.4	(1-3)x10^{-4}
SCF	Pc/Tc	200-500	0.7x10^{-3}	(1-3)x10^{-4}
Liquid	0.1/298	600-1600	2x10^{-6}-2x10^{-5}	(0.2-3)x10^{-2}

Table 2.2 presents several substances which may be used as supercritical fluids. Especial attention is given to carbon dioxide (CO_2), the most commonly used supercritical fluid worldwide. In comparison to other substances, CO_2 presents a critical temperature (Tc) close to room temperature (31.1 °C) and a relative low critical pressure (Pc=7.38 MPa), what is interesting when considering the energy requirements for the solvent delivery at a determined operational pressure. Besides, it is abundant, relatively inexpensive, inert, can be used in high purity, non-flammable, atoxic and non-explosive, following the environmental and health organizations´ restrictions [2].

Table 2.2. Critical data of some pure components [1].

Component	Tc (°C)	Pc (MPa)
Ethylene	9.4	5.04
Carbon dioxide	31.1	7.38
Ethane	32.3	4.87
Nitrous oxide	36.6	7.26
Propane	96.8	4.25
n-Hexane	234.5	3.01
Acetone	235.1	4.70
Methanol	239.6	8.09
Ethanol	240.9	6.14
Ethyl acetate	250.2	3.83
Water	374.1	22.06

Variations in pressure and temperature can be used to adjust the solubility of a substance in supercritical carbon dioxide (SC-CO_2). Due to its compressibility, a pressure increase will result in a higher density, which leads to more solute-solvent interactions, i.e., mixing and penetrating small pores in a solid matrix more effectively. Figure 2.2 shows the CO_2 density as a function of temperature and pressure. Close to the critical point, a slight change in the operational conditions (pressure and temperature) may cause a drastic variation in its density, affecting consequently the solubility of the solute in the supercritical phase [1].

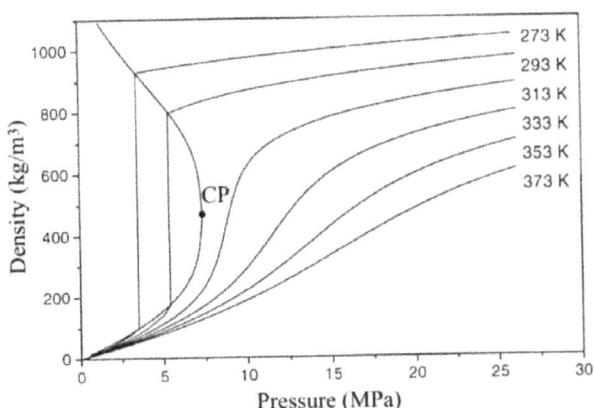

Figure 2.2. Density of pure CO_2 at different conditions.

The solvent power is also influenced by the polarity of the fluid. The solubility of chemical substances in non-polar solvents decreases with increasing molecular weight and mostly with increasing polarity and number of polar functional groups. CO_2 is a non-polar solvent, what makes it unable to dissolve polar substances. In order to increase the efficiency of the process through the enhancement of the solvent's polarity, the addition of small percentages of modifiers or entrainers (liquid substances, like ethanol and other alcohols) or co-solvents (substances in gaseous phase, like ethane and propane) can be employed [2]. However, the separation of modifiers from the final products may consist a problem, especially when a high purity of the products is required, since they remain as solvent residues when the process is finished.

2.1.1. Basic Characteristics of Gas Extraction Processes

2.1.1.1. Supercritical Fluid Extraction

Supercritical Fluid Extraction (SFE) from solid particles consists one of the most widely used applications of SCF. The extraction is carried out through the continuous contact between a solid matrix and the solvent at high pressures. The solid substrate is loaded into the extractor, forming a fixed bed of particles, and the SCF flows through it (see also Chapter 2.5), allowing the solubilization of the desired components. The desired components are then continuously extracted by the supercritical phase until the solid matrix is depleted.

The process consists basically of two steps: the extraction itself, and the separation between extracted components and the solvent. A simplified flowsheet can be observed in Figure 2.3. Once the fixed bed is formed, the solvent is fed at the desired pressure and temperature conditions, flowing through the loaded extractor. The loaded (or even saturated) solvent is then removed to the separator in order to perform the separation solute/solvent by means of an increase in temperature or a drastic pressure reduction. If the extraction process is conducted at smaller equipments (laboratory or pilot scale ones), the regenerated gas phase can be released to the atmosphere. Otherwise, if the gas consumption is high (leading consequently to higher solvent costs) or if it is considered flammable, explosive or toxic, the gas can be recycled and fed once more into the extractor, closing the solvent cycle and avoiding the discharge of the solvent in the atmosphere.

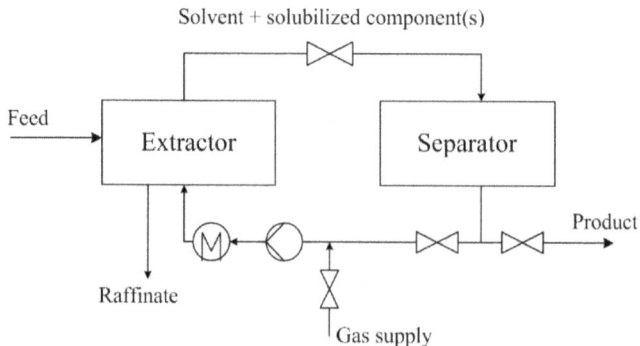

Figure 2.3. Flowsheet of SFE from solid materials (adapted from [1]).

2.1.1.1.1. Course of Extraction for SFE from Solid Materials

As previously presented, the extraction of components from solid materials is performed through the contact between the solid matrix and a continuous solvent flow. During the process, the concentration of solute in the fluid phase increases, while the amount of extractible components in the solid phase decreases. The concentration of these components in the interior of each solid particle decreases slower than at the surface, providing then the concentration gradients necessary for the extraction process. The concentration variations during the process are dependent on the extraction kinetics, properties of the solid material and the solvent capacity of the SCF, what can be changed according to the operational conditions employed.

The course of a solid extraction can be represented by the overall extraction curves (OEC), where the amounts of extract collected during the process are plotted as a function of time or as a function of the amount of solvent used [1, 2]. Figure 2.4 shows two schematical extraction curves. The first part of both curves (I) is linear, corresponding to a constant extraction rate, where the amount of solute collected remains constant and the mass transfer is mostly influenced by convective effects. The concentration gradients in the linear portion of the OEC may be represented by the equilibrium solubility, what can be assumed when using low solvent flows, achieving then the saturation of the solvent with the solute. The second phase (II) represents the falling extraction period, characterized by the decrease of the extraction rate. Both effects, convection and diffusion in the solid phase are responsible for the determination of the extraction rate. Finally, the third phase (III), which is only shown for curve (a), corresponds to the end of the process, where the solid material is depleted and the diffusive phenomena in the solid phase are predominant. The maximum extraction yield is then limited by the total amount of extractible substances present in the matrix, represented by the dashed line in Figure 2.4.

Figure 2.5 shows the course of the extraction from solid materials for the extraction rate, providing very useful information about the process. As presented by Brunner [1], two types of extraction curves can be evaluated. Curve 1 represents the amount of extract collected per unit of time of a high initial concentration of extract in the solid matrix. A good example are oilseeds, because they present an extract readily accessible for the solvent. In the beginning of the extraction, mass transfer is constant and determined by the different concentrations at the interface between solid and fluid phases. After the initial period, the decrease in the medium concentration of extract in the solvent can be due to an additional transport resistance, which is caused by the depletion of the extract in the solid matrix near the interface gas/solid. Additionally, the solvent may leave the extractor unsaturated, mainly because the extractor is not long enough to enable the solvent saturation.

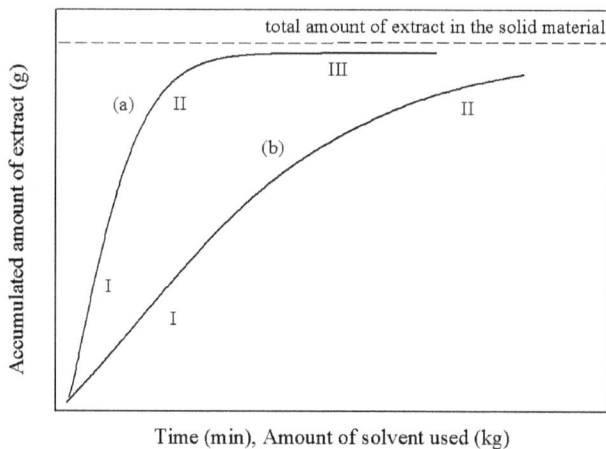

Figure 2.4. Overall extraction curves obtained through SFE from solid materials.

The extraction rate for cases with a low initial concentration of extract in the solid matrix or an extract not readily accessible (transport within the solid to the interface prevails since the beginning of the experiment) is represented by curve 2. Coffee beans are a good example. From Figure 2.5, it can be seen that curve 2 corresponds to the second part of curve 1, because a depletion phase always follows the first extraction phase of constant concentration downstream the extractor.

Additionally, the process may exhibit a region where the extraction rate increases with time. This usually happens when starting up the extraction apparatus. If solvent ratios are relevant for technical purposes, this initial part may be small or even considered negligible. But at lower solvent flows, it may extend over a substantial part of the extraction and the analysis of the results must be carefully performed.

There are many different aspects to be evaluated when aiming the modeling of the OEC. The extraction can be evaluated from the point of a single particle, from the point of a fixed bed and also from the solvent. The mathematical modeling of the OEC will be presented in Chapter 2.5.

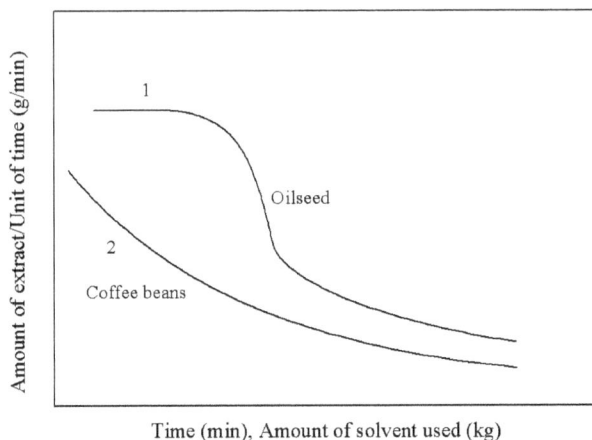

Figure 2.5. Extraction rate curve for two different solid materials. Operational conditions and properties of the solids are identical (adapted from [1]).

2.1.1.2. Countercurrent Multistage Extraction

If the yield or selectivity of an extraction using a simple one-stage apparatus is not sufficient, the liquid feed and supercritical solvent can be inserted countercurrently into one separation apparatus. Such a separation column will enable multiple equilibrium stages through the enhancement of the mass-transfer area between gas and liquid phases by using structured packings, improving consequently the process efficiency.

A simplified flowsheet of a countercurrent multistage extraction process is presented in Figure 2.6. The equipment is generally composed by a packed separation column, a separator vessel responsible for the solvent-extract separation, feed and reflux pumps, a solvent pump or compressor (depending on the physical state of the solvent), and auxiliary devices for recovering top and bottom products (extract and raffinate fractions, respectively).

The process is in many aspects comparable to rectification and it can be operated with or without reflux. Reflux must be employed in order to provide the enrichment of the volatile components in the extract fractions (top product). When employing extract reflux, a stripping section will be localized below the liquid phase feed inlet, where the top products (high volatile components) are separated from the bottom products and then transported to the enriching section. The enriching section is located between the liquid feed and extract reflux inlet points, where the bottom products (low volatile components) are separated from the top product components and transported to the lower section of the column (stripping section).

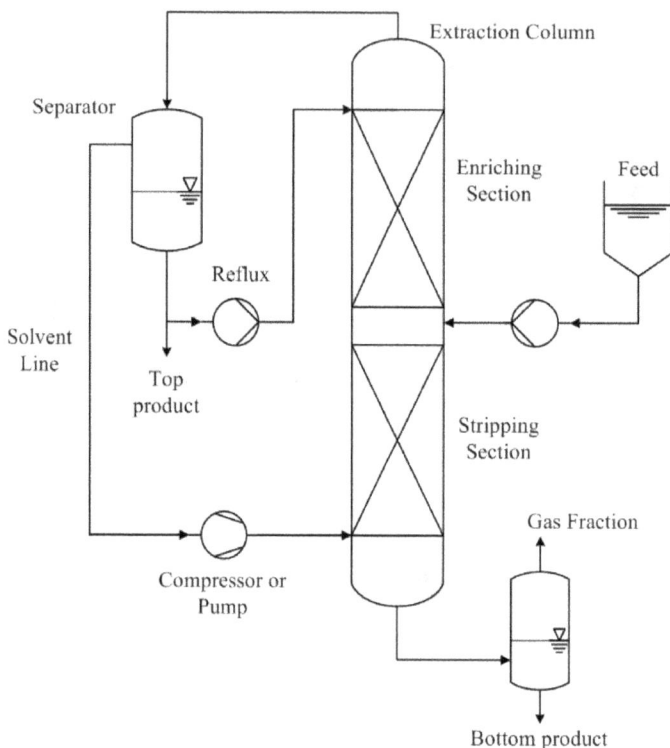

Figure 2.6. Simplified process scheme of a countercurrent multistage extraction apparatus.

In order to evaluate the process efficiency, the number of components to be fractionated must be carefully analyzed. The separation of two components into practically pure substances is possible, representing a basic case. In fact, this case is very unusual in practice, since multicomponent complex mixtures are often separated with multistage countercurrent gas extraction. When several desired components with similar properties are present in the feed material, they can be treated as a pseudocomponent, considering then the feed-mixture as a quasi-binary system. Alternatively to the pseudocomponent approach, two key-components which are hardest to separate can be considered, assuming that the separation of the other components takes place simultaneously.

2.1.2. Applications of SCF – Overview

Since Baron Cagniard de la Tour observed the occurrence of a supercritical phase [3], compressed gases have been employed in order to dissolve substances of low volatility. In 1879, Hannay and Hogarth [4] investigated the solubility of metal chlorides in supercritical ethanol. However, high pressure gases were employed for industrial purposes only after the middle 1930's: in 1936 Wilson et al. [5] designed a process for the deasphalting of lubricant oils using propane at near critical conditions. Propane was also used as solvent in the so-called Solexol process, in order to concentrate polyunsaturated triglycerides of vegetable oils [6] and extract vitamin A from fish oils.

The first large scale SFE processes for food industry applications, such as caffeine extraction from coffee beans (usually performed with dichloromethane) and tea, and aroma and flavor extraction from hops, were developed in Germany. Tobacco nicotine removal was firstly mentioned in the early 70's. For these applications, the industrial plant capacities reached approxim. 64,000 liters volume up to year 2004 [7].

During the last 2 decades, a large number of industrials plants (approx. 100) of different volume sizes were built for batch extractions of solid materials with SCF. In this time, a total number of about 100 extractor vessels larger than 100 liters have been designed for different industrial plants, distributed mainly in Europe, Japan, USA and in other Asian countries [8]. Table 2.3 summarizes some applications of SCF in pilot and industrial scales around the world [7].

The DASFAF ("Developments and Applications in Supercritical Fluids in Agriculture and Fisheries") Network presented one state of the art book on supercritical fluids, providing a general overview of the actual and future trends in high pressure processes. The topics included several potential applications of SCF: phase equilibrium measurements and modeling, adsorption processes, generation of micro- and nanoparticles, soil and waste treatment cleaning, separation of extracts with membranes, aerogels, dyeing processes, development of polymers in supercritical mediums, recovery of polishing earths and reactions of natural products in superheated water. Special attention was given to SFE and fractionation of solid and liquid natural materials, as well as to industrial developments and economic issues involved.

Table 2.3. Main applications of SCF in pilot and industrial scales (up to 2004) [7].

Application	Solvent	Stage of development
Coffee and tea decaffeination	CO_2	Industrial plants (USA and Germany)
Hop resins	CO_2	Industrial units (Australia, UK, Germany, USA)
Spices	CO_2	Industrial units (France, Germany, Switzerland and Japan)
Aromas/essential oils	CO_2	Laboratory and pilot scales
Vegetal oils	CO_2	Pilot plants
Food fat removal	CO_2	Pilot plants
Nicotine removal	CO_2	Industrial plants (USA)
Medicinal plants	CO_2	Industrial (Japan) and pilot (Germany) scales
Earth decontamination	CO_2	Pilot plants
Mud treatment	Propane	Industrial plants (USA)

In the field of SFE, supercritical extraction and fractionation of natural matter are probably the most studied applications. In the last 15-20 years, several studies on the extraction of classical components from plant materials (seeds, fruits, leaves, flowers and rhizomes), including antioxidants, pharmaceuticals, coloring matters and pesticides, with or without the addition of a co-solvent have been published. Several reviews summarizing these works on SFE and fractionation of natural products have been published in the last years [8-12].

Essential and citrus oil deterpenation through SCF treatment was firstly suggested by Stahl et al. [13] and since then the number of publications has increased considerably. The same tendency has been observed for the extraction/refining of edible oils. More information on the state of the art of the application of SCF in the extraction/fractionation of rice bran and mandarin peel (citrus) oils are presented and discussed in details in Chapters 3.3 and 4.3, respectively.

2.2. Phase Equilibria – Fundamentals

Phase equilibria is the most important basis for understanding the phenomena concerned with several separation processes, especially gas extraction. Since the thermodynamic equilibrium is achieved, it can provide valuable information about the dissolution capacity of a determined solvent, the amount of solvent dissolved in the liquid phase and the equilibrium composition of the liquid phase, the dependence of the solvent properties with operational conditions, the extent of the two-phase area and the solvent selectivity [1, 2].

In a system composed by two or more components, two or more phases may coexist in equilibrium. The equilibrium is assumed when temperature (T), pressure (P) and chemical potential (μ) of one system composed by n components and π coexisting phases are equalized (Equations 2.1-2.3). In order to differentiate the compositions in each phase, y_i is denoted as the concentration of component i in the gaseous phase, while x_i denotes its concentration in the liquid phase [14].

$$T^{(1)} = T^{(2)} = \ldots = T^{(\pi)} \tag{2.1}$$

$$P^{(1)} = P^{(2)} = \ldots = P^{(\pi)} \tag{2.2}$$

$$\mu^{(1)} = \mu^{(2)} = \ldots = \mu^{(\pi)} \tag{2.3}$$

When the system temperature is constant, the fugacity of a pure substance (f_i) may be considered a corrected partial pressure. Then, the chemical equilibrium can be presented as follows (Equation 2.4):

$$f_i^{(1)} = f_i^{(2)} = \ldots = f_i^{(\pi)} \tag{2.4}$$

In order to evaluate the feasibility of a separation process, the system must be well characterized thermodynamically. The first step is related to the number of components to be considered for one determined separation task. In the next sections, a brief introduction on phase equilibria of binary and ternary systems will be presented. Further discussions can be found in several text books, including also discussions on quaternary systems [1, 2].

2.2.1. Binary Systems

Phase equilibrium data are usually represented graphically. A three-dimensional diagram (P-T-x) of a binary system composed by a supercritical component and a component with medium volatility is presented in Figure 2.7. When reaching the limiting composition values for each component (x=0 and x=1), vapor pressure curves of both components are responsible for separating gas from liquid regions, finishing then at the critical point of each component (CP$_A$ and CP$_B$).

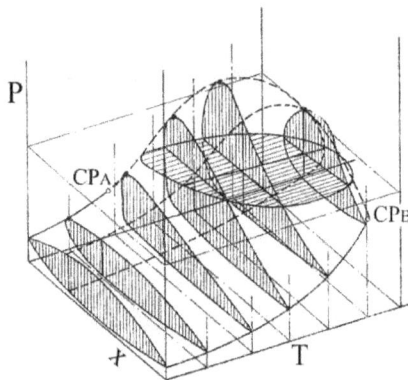

Figure 2.7. P-T-x diagram of one binary system (adapted from [1]).

The critical curve connects the presented critical points. Between the vapor pressure curves of both components and below the critical curve (represented by dotted lines), the two-phase region is located. Depending on the phase behavior of a system, the critical curve can present different shapes, assuming up to 6 different forms [15]. The behavior presented by Figure 2.7 is called as type I and is the most simple one.

In order to better visualize the phenomena, two-dimensional diagrams can be plotted by cutting the original three-dimensional diagrams: at constant temperatures (P-x diagrams), at constant pressures (T-x diagrams) and at constant compositions (P-T diagrams). A P-x diagram is presented in Figure 2.8.

In a P-x diagram, composition is plotted in the abscissa and the pressure in the ordinate axis. The equilibrium compositions of coexisting phases are plotted at determined temperature condition. If the pressure of the system varies isothermically, the composition of coexisting phases form one closed loop. At a pressure P_1, a liquid phase L_1 and a gas phase V_1 coexist in equilibrium for a determined composition x_1. The line L_1V_1 is a tie line, which connects coexisting equilibrium phases.

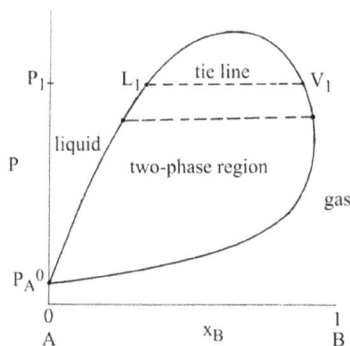

Figure 2.8. P-x diagram for a binary mixture.

The closed loop encloses the two-phase area. The compositions of both coexisting phases begin from the vapor pressure of component A ($P_A{}^0$), the subcritical component, converging then to the critical point, where liquid and gas phases are identical. Since component B is supercritical, the component presents no vapor pressure at T_1. At pressures higher than the vapor pressure of A, a homogeneous liquid mixture of A and B is formed; at pressures below its vapor pressure, homogeneous gas mixtures of A and B exist. Finally, at pressures higher than the critical pressure, both components are completely miscible.

2.2.2. Ternary Systems

In a three component system commonly employed in gas extraction processes, one component is always the supercritical fluid. The other two components to be separated are normally one high volatile component (HVC) and one low volatile component (LVC). For multicomponent mixtures, the assumption of pseudo-components may be useful, since the mixture can be then treated as a pseudo-ternary system.

A ternary system composed by a supercritical fluid (CO_2), a low volatile component and a high volatile component is very often represented in triangular plot diagrams, the Gibb´s diagrams, as illustrated in Figure 2.9. The operating pressure and temperature are kept constant. Each corner of the diagram corresponds to one pure substance. Binary systems are represented by the side lines connecting two pure components. As noticed by analyzing these diagrams, LVC and HVC are miscible.

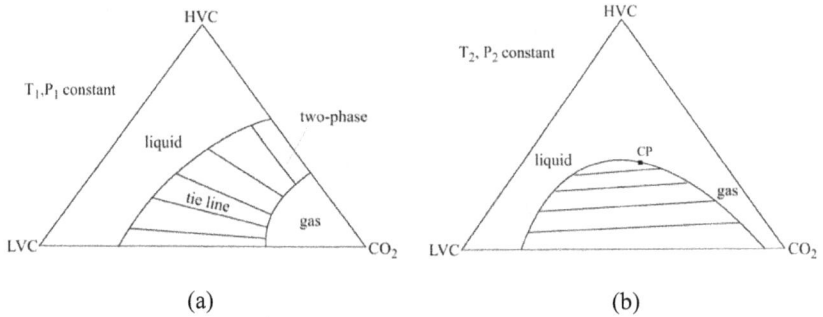

(a)   (b)

Figure 2.9. Phase equilibrium for a ternary system CO_2-LVC-HVC ($P_1 < P_2$).

At lower pressures (Figure 2.9-a), both components are partially miscible with the SCF. The two-phase region is formed and the gas and liquid phases present different compositions, what is highly desired for every separation task. The coexisting phases are connected to each other by their respective tie lines. At higher pressures (Figure 2.9-b), a closed miscibility gap is formed and the HVC becomes then completely miscible in the supercritical gas. Consequently, the range of the two-phase area is reduced. This effect can also be obtained when employing modifiers, which ones have to be chosen accordingly to the operational conditions employed and the system investigated [1].

2.3. Modeling Multistage Countercurrent Processes

In order to improve the separation efficiency of countercurrent processes, fractionation of liquid materials is usually carried out in packed columns, improving the contact between both phases and, consequently, the mass transfer rates [1].

Different methods are available for modeling multistage extraction processes. For stage calculations, the concept of theoretical stages is mostly employed. The column is divided then in different sections, where phase equilibrium between the phases involved is achieved. Masses are transferred from one stage to the others, forming a cascade of equilibrium cells. The process is schematically shown in Figure 2.10.

The operating lines for the enriching and stripping sections are determined by performing a material balance (solvent-free-basis) on the upper and lower sections of the multistage cascade process. Assuming that the process is carried out at constant pressure and temperature, it can be written for the enriching section (Equations 2.5 and 2.6):

$$V_p y_{1_p} - L_{p+1} x_{1_{p+1}} = V_n y_{1n} - R_n x_{1Rn} \tag{2.5}$$

and

$$V_p - L_{p+1} = V_n - R_n \tag{2.6}$$

For the stripping section, the following mass balances are (Equations 2.7 and 2.8):

$$V'_{p'} y_{1_{p'}} - L'_{p'+1} x_{1_{p'+1}} = S_0 y_{1S_0} - L'_1 x_{11} \tag{2.7}$$

and

$$V'_{p'} - L'_{p'+1} = S_0 - L'_1 \tag{2.8}$$

The total mass balance is then:

$$F + S_0 + R_n = V_n + L'_1 \tag{2.9}$$

where:

y_1, x_1 = concentrations of component 1 in the gas and liquid phase
y_{1p}, x_{1p} = concentrations of component 1 in the gaseous and liquid flow from stage p
V_p = gaseous flow from stage p
L_p = liquid flow from stage p
R_n = reflux to stage n of the enriching section
S_0 = solvent to the first stage of the stripping section
F = feed mixture
L'_1 = liquid flow leaving stripping section (raffinate)
V_n = gas flow leaving enriching section (extract)

The above set of equations can also be implemented into computational programs. Some commercial simulators, such as ASPEN+®, HySIS® and PROII® present special routines for the calculations of multistage separation processes, including also application to special industrial cases.

The literature presents some short-cut methods for the determination of the number of theoretical stages through graphical estimatives and numerical calculations. The methods to be mentioned are the McCabe-Thiele and the Ponchon-Savarit ones. Further discussions about these methods are presented by Treybal [16].

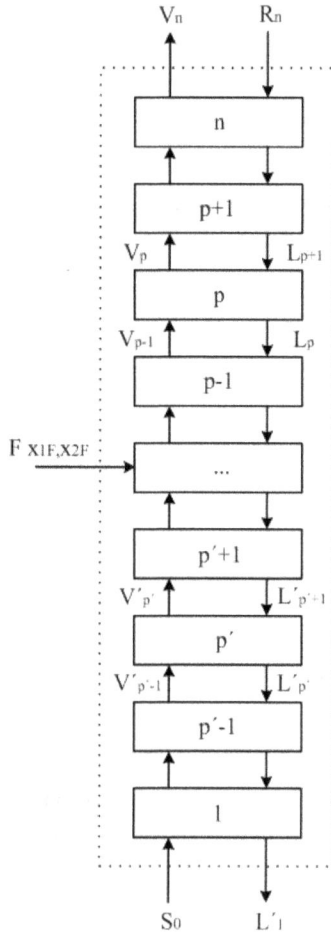

Figure 2.10. Flowsheet of a multistage separation process for a binary mixture.

The McCabe-Thiele diagram illustrates the multistage separation of a binary mixture and is presented in Figure 2.11. Phase equilibrium and material balances are plotted in a two-dimensional diagram, where the ordinate values represent concentrations of component A in the gas phase, while the abscissa values represent compositions of component A in the liquid phase. Equilibrium can be represented if the separation factor is considered constant, otherwise the definition of the equilibrium line must be done by experimental points or by correlating functions based on experimental data.

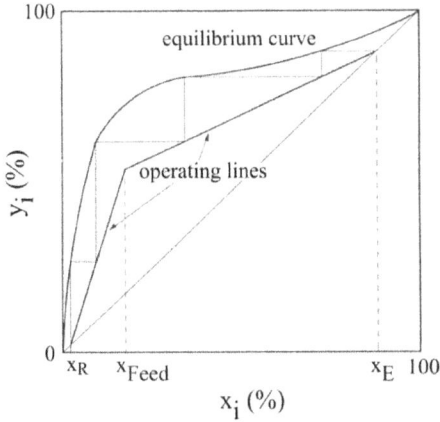

Figure 2.11. McCabe-Thiele diagram.

The slope of the operating lines can be influenced by reflux and solvent-to-feed ratios, and mutual solubility of liquid and solvent mixtures. If the mutual solubility remains constant , the flows of each phase in both sections of the column remain constant. In this case, the operating lines derived from mass balances are straight. An infinite number of stages can be achieved at minimum reflux ratio and the minimum number of stages can be reached when employing maximal (infinite) reflux ratios [1].

The mass balances are linear equations, due to the fact that V and L are constant. The operating line for the enriching section is represented by Equation 2.10:

$$y_A = \frac{V}{V+1}x_A + \frac{1}{V+1}x_E \tag{2.10}$$

For the stripping section (Equation 2.11):

$$y_A = \frac{L}{V}x_A - \frac{L}{V}x_R + y_0 \tag{2.11}$$

with:

y_A= concentration of the volatile phase A in the vapor phase
x_A= concentration of A in the liquid phase
x_R= concentration of A in the raffinate
x_E= concentration of A in the extract
y_0= concentration of A entering with the solvent

L = liquid mass flow
V = vapor mass flow
R = raffinate mass flow
E = extract mass flow
v = R/E = reflux ratio

The liquid flow changes at the feed point due to the combination between feed and reflux flows. So, parameter L in the stripping section corresponds to its total liquid flow. A common simplification corresponds also to the assumption of saturation of solvent with part of the raffinate at the bottom of the column ($y=x_R$). As a final consideration, concentration changes in one theoretical stage may be large, causing a substantial variation on the flowing phases. Then, it could be necessary to apply the Ponchon-Savarit method in order to provide a better representation of the process.

The Ponchon-Savarit method is applicable for the separation of complex mixtures, presenting the possibility to reduce the system to an equivalent ternary system composed by two key-components and the supercritical fluid itself. Additionally, the mutual solubility of the phases and the separation factor may vary with the concentration [1]. With these considerations, the method can overcome the restrictions of the McCabe-Thiele method.

This equivalent mixture can then be represented in a rectangular diagram, where the solvent ratio (ordinate) is plotted against the concentration of one of the key-components (abscissa) in a Jänecke diagram, as presented in Figure 2.12. The upper line represents the gas phase (extract line), and the lower one the liquid phase (raffinate line). In most practical cases the liquid line will coincide with the abscissa because the solvent ratio in the liquid phase is on the order of 10^0.

The mass balance for component 1 in the enriching section can be defined as:

$$V_p y_{1_p} - L_{p+1} x_{1_{p+1}} = V_n y_{1n} - R_n x_{1Rn} = E y_{1E} \qquad (2.12)$$

where E corresponds to the amount of extract (including solvent) collected at the top of the column and y_{1E} the concentration of component 1 in the extract.

Since it is assumed that the process achieves steady-state conditions ($E y_{1E}$ is constant), all lines representing mass balances within this column section will converge to one single point, P_E, defined by $E y_{1E}$. The amount of solvent is given by $V_n y_{Sn} - R_n x_{SRn}$ and the mass balance for the stripping section is presented in Equation 2.13:

$$V'_{p'} y_{1_{p'}} - L'_{p'+1} x_{1_{p'+1}} = S_0 y_{1S_0} - L'_1 x_{11} = B x_{1B} \qquad (2.13)$$

where B corresponds to the bottom product (including the solvent).

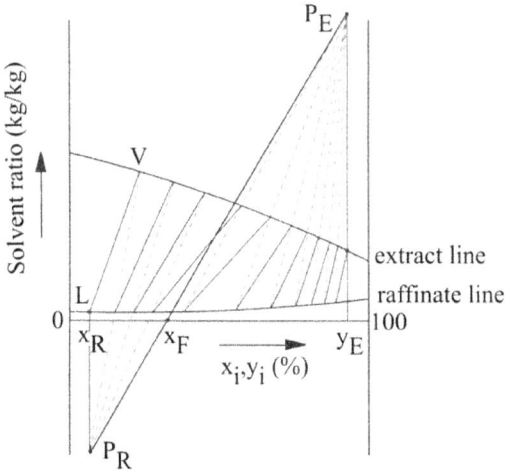

Figure 2.12. Determination of the number of stages by a Jänecke diagram with Ponchon-Savarit method.

Analogically to the enriching section, all lines representing mass balances within the section will converge to one point, P_B, defined by Bx_{1B}; the amount of solvent is then defined as $S_0 y_{SS0}-L_1{'}x_{S1}$. The total mass balance is given by $F=E+B$. More information about the calculations, including a solution procedure for a separation task are provided by Brunner [1].

The applicability and efficiency of both methods listed here depends on reliable phase equilibrium data. When covering a wide spectrum of concentrations (from 0 to 100 wt.-%) and separation factors, the curvature at the end points of the Jänecke diagram can be better determined, since they can be associated with a steep increase in the number of theoretical stages for one separation task, what will affect the assembling and operational costs involved.

2.4. Principles on Ad- and Desorption Processes

Adsorption occurs whenever a solid surface is exposed to a gas or a liquid. It is basically defined as the enrichment of material or increase in the density of the fluid in the vicinity of an interface [17]. Under certain conditions, highly porous and very fine media are preferred as adsorbent, since they can provide larger specific surface areas. Usual examples are activated carbon, silica gel, clays, zeolites and aluminophosphates, among others [18].

Adsorption is a very important process in the industry: some adsorbents are used in large amounts as desiccants, catalysts or catalysts supports; others are used for the separation of gases, liquid purification and pollution control. Additionally, adsorption processes play an important role in solid state reactions and biological mechanisms [18].

The reverse process to adsorption is called desorption. The adsorptive is initially present in the fluid phase and afterwards accumulated on the adsorbent as adsorbate. Figure 2.13 represents the ad- and desorption phenomena at the adsorbent's interface.

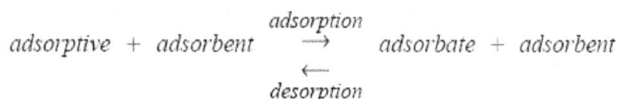

$$\textit{adsorptive} \; + \; \textit{adsorbent} \; \overset{\textit{adsorption}}{\underset{\textit{desorption}}{\overset{\longrightarrow}{\longleftarrow}}} \; \textit{adsorbate} \; + \; \textit{adsorbent}$$

Figure 2.13. Representation of ad- and desorption processes at adsorbent's interface.

Adsorption at a surface or interface is caused largely by binding forces between the individual atoms, ions, or molecules of an adsorbate and the surface. Adsorption is an exothermic process. Physical adsorption is mainly the result of van der Waals and electrostatic forces between adsorbate molecules and the atoms that compose the adsorbent surface. Because of the small range of these forces, the adsorbent and adsorbate have a loose bond, which can be easily released [18].

The adsorption enthalpies lie between the heat of condensation and the enthalpy of a chemical reaction. Several factors influence the adsorption of a certain compound to an adsorptive. For the adsorptive, they include molecular weight and size, structure, polarity, and its affinity to the fluid phase. For the adsorbent, the chemical composition of the surface, pore size distribution and their configuration, as well as size and form of the particles determine the adsorption properties. Furthermore system conditions like pH, temperature, and pressure affect the adsorption [17, 18].

Desorption of the adsorbed solute from the adsorbent is accomplished in one or two general methods. The first method consists of changing the physical operational conditions in order to lower the equilibrium interaction between the adsorbent and the solute (adsorptive). This could range from passing a different feed stream through the adsorbent to decreasing the pressure of the system. The second method corresponds to the modification of the nature of the adsorbed material by carrying out chemical reactions. Then, the adsorbed material can be desorbed and removed from the system readily. This method could range from passing a feed solution, which reacts with the adsorbed material to thermal type regenerations at high temperatures where under a controlled gaseous atmosphere, preferred chemical reactions occur with the adsorbed components [18].

The relation between the adsorptive's concentration in the fluid phase and its loading on the adsorbent is described by adsorption isotherms. These curves are obtained by measuring the equilibrium at different concentrations and constant temperatures. They can be described by different models. For instance, the Langmuir-approach assumes a constant adsorption energy, no interaction among adsorbates, and monomolecular coverage of the surface [18, 19]. The rate of adsorption is proportional to the fraction of uncovered surface while the rate of desorption is proportional to the degree of coverage. The Langmuir-isotherm can be represented by Equation 2.14 [20]:

$$q_e = \frac{Q_0 b C_e}{1 + b C_e} \tag{2.14}$$

where:
q_e = equilibrium loading
Q_0 = loading at full coverage
C_e = equilibrium concentration in the solvent
b = constant of the model

Adsorption and following selective desorption was reported as feasible for the deterpenation of citrus peel oils, what will be discussed further (see Chapter 4.3.1). In this process, the oil is adsorbed and selectively desorbed with supercritical fluids at varying conditions similar to solid extraction. Alternatively, only the components with a higher affinity to the adsorbent are adsorbed, either by mixing an excess of oil with the adsorbent or by dissolving all the oil in the fluid and bringing it in contact with the adsorbent at conditions similar to the first extraction step during selective desorption.

The method makes use of the different adsorbing power of the the oil components to the active sites on the adsorbent surface. Silica gel is a very versatile adsorbent. About 80 % of its surface is constituted by polar heads via unreacted or residual silanols (see Figure 2.14). The other 20 % are nonpolar methyl groups [21].

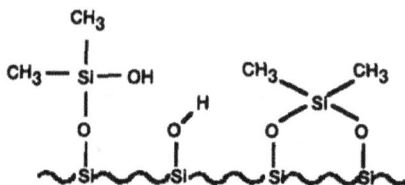

Figure 2.14. Schematic structure of silica gel surface [21].

Nonpolar components can be only adsorbed by these nonpolar groups, while polar compounds are adsorbed by the polar sites. Furthermore, the molar mass of the substance contributes to the desorption speed. The course of desorption is similar to an extraction from a solid matrix: the accumulated yield curve shows a steep linear increase in the beginning of the process. When the remaining mass of adsorbate on the adsorbent is not readily available anymore, the slope decreases as the mass-transfer rate is limited. Finally, the accumulated mass approaches a constant level when the adsorbate is depleted, achieving then the end of the process.

2.5. Mathematical Modeling of Solid Extraction Processes

In the last decades, great advances in the SFE of organic compounds from several plant materials have been achieved, mostly due to the necessity of producing high quality extracts in a solvent-free basis.

In order to optimize the operational conditions of SFE processes (achieving higher extraction yields), mathematical models should be employed. Process design parameters, such as equipment dimensions, solvent flow rates and particle diameters, must be carefully evaluated, intending to predict and extend the viability of a determined SFE process from laboratory to pilot and industrial scales through the simulation of the overall extraction curves (OEC). However, a model should be employed not only as a simple mathematical instrument; it should really reflect the physical phenomena involved in the dissolution of the desired substances, taking into account aspects related to the structure of the vegetable matrix, as well as previously experimental observations for each particular case [1, 10].

There are several mathematical models in the literature to fit SFE extraction curves to experimental data. Some of these are totally empirical, but the majority is based on mass balances within the solid and fluid phases [1, 10, 22-25]. Most of the models available treat the obtained extract as a pure substance (the so-called pseudocomponent), although it may contain several components from different chemical classes.

According to del Valle and de la Fuente [26], most interesting mass transfer models are the ones based on mass balance equations for infinitesimal sections of a solid packed bed: in order to mathematically describe the process, a cylindrical extraction vessel (length L, inner diameter d_E) can be divided into finite difference volume elements (height Δz), as can be visualized in Figure 2.15. Mass balances must be then written and arranged when taking some boundary conditions into account, like $\Delta z \rightarrow 0$ and $\Delta t \rightarrow 0$.

For a brief overview of some models already presented in the literature, the model proposed by Akgerman and Madras [27] will be used as a general model. The model is summarized in Table 2.4 and the assumptions made are presented as follows:

- the solid particles are treated as spherical ones, porous and homogenous;
- interstitial velocity of the fluid is constant, i.e., bed porosity (ε) remains unchanged;
- physical properties of the SCF and substrate in the bed are constant;
- pressure losses and temperature gradients are negligible within the bed;
- axial dispersion of the solute in the SCF .

In relation to the last assumption, dissolved solute in the SCF flows along the bed (diffusion coefficient D_L) from high concentration zones to low concentration zones.

In the solid matrix two solute fractions can be found, the solute is adsorbed into the solid (concentration C_s), and the solute dissolved in the fluid phase within the pores (concentration C_p). Initial concentrations C_{s0} and C_{p0} are determined by the total initial solute concentration (C_0) and equilibrium between the two phases, Equation 2.15:

$$C_{s0} = f(C_{p0}) \tag{2.15}$$

$$\varepsilon_p C_{p0} + \rho_p C_{s0} = C_0 \tag{2.16}$$

By assuming the considerations above, solute concentrations in the solid matrix (C_s and C_p) depend on radial position within the particle (r), axial position (z), and extraction time (t); whereas the total solute concentration in the SCF (C_f) depends only on z and t.

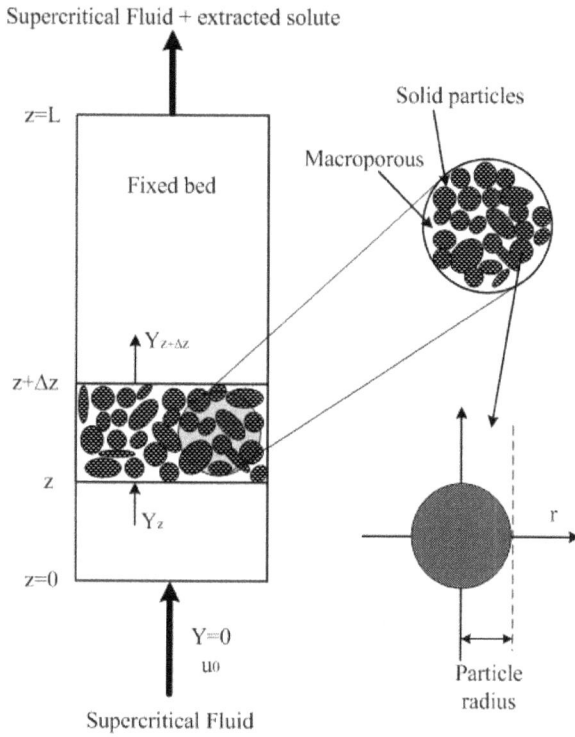

Figure 2.15. Scheme of a packed bed extractor and a difference volume element.

Table 2.4. General model for SFE of solid matrixes [26].

Differential mass transfer in the bulk fluid phase:

$$\frac{\partial C_f}{\partial t}+u\frac{\partial C_f}{\partial z}-D_L\frac{\partial^2 C_f}{\partial z^2}=\frac{3}{R}\frac{1-\varepsilon}{\varepsilon}J \qquad (2.17)$$

Initial condition: $C_f=0$	(at $t=0$, for all z)	(2.18)

Boundary conditions: $D_L\dfrac{\partial C_f}{\partial z}=uC_f$ (at $z=0$, for all t) (2.19)

$$\frac{\partial C_f}{\partial z}=0 \qquad \text{(at } z=L, \text{ for all t)} \qquad (2.20)$$

Differential mass transfer in the solid phase:

$$\frac{\partial C_p}{\partial t}+\frac{\rho_p}{\varepsilon_p}\frac{\partial C_s}{\partial t}=\frac{D_e}{r^2}\frac{\partial}{\partial r}\left(r^2\frac{\partial C_p}{\partial r}\right) \qquad (2.21)$$

Initial conditions: $C_p=C_{p_0}$ (at $t=0$, for all r and z) (2.22)

$C_s=C_{s_0}$ (at $t=0$, for all r and z) (2.23)

Boundary conditions: $\dfrac{\partial C_p}{\partial r}=0$ (at $r=0$, for all z and t) (2.24)

$\dfrac{\partial C_s}{\partial r}=0$ (at $r=0$, for all z and t) (2.25)

$\varepsilon_p D_e\dfrac{\partial C_p}{\partial r}=-J$ (at $r=R$, for all z and t) (2.26)

Definitions:

Mass Transfer rate: $J=k_f(C_p|_R-C_f)$ (2.27)

Equilibrium relationship: $C_s=f(C_p)$ (2.28)

The model proposed by Crank [28] is based on the considerations of a solid sphere particle (radius r), presenting an uniform initial concentration of dissolved material, which one is immersed into a fluid. It is the necessary to solve the diffusion equation for the system with appropriate boundary conditions. In this case, the problem can be mathematically solved in a similar way to that of the immersion of a hot sphere into a cold fluid. Therefore, it can be defined as a hot-ball model. Adaptations of the published solutions led to the following equation for the mass ratio m, the extractable material remaining in the matrix sphere after time extraction t, and for the initial extractable material mass m_0 [28, 29]:

$$\frac{m}{m_0}=\frac{6}{\pi^2}\sum_{n=1}^{\infty}\frac{1}{n^2}\exp\left(\frac{-n^2\pi Dt}{r^2}\right) \qquad (2.29)$$

where n is an integer and D the diffusion coefficient of the material in the sphere.

Tan and Liu [25] presented a model for desorption of ethyl acetate through the supercritical regeneration of activated carbon. The model assumes no axial dispersion and approximates the desorption kinetics as being linearly related to the adsorbed concentration. It was used to represent the OEC of curcumins extraction from *Curcuma longa* L. [30] and in the SFE of marigold oleoresin [31], with good agreement between experimental and correlated results.

The mass balance in the bulk phase in the column may be written by (Equation 2.30):

$$\varepsilon \frac{\partial C}{\partial t} + u \frac{\partial C}{\partial z} = -(1-\varepsilon)\frac{\partial S}{\partial t} \qquad (2.30)$$

The initial and boundary conditions are:

$$t = 0, \quad C = 0 \qquad (2.31)$$
$$z = 0, \quad C = 0 \qquad (2.32)$$

The mass balance in the activated carbon is expressed in terms of linear desorption kinetics which may be written by Equation 2.33:

$$\frac{\partial S}{\partial t} = -kS \qquad (2.33)$$

where k is defined as the kinetic desorption constant. The initial condition is:

$$t = 0, \quad S = S_0 \qquad (2.34)$$

The concentration at the exit of the column can then be obtained by solving the above equations and is expressed by:

$$C_e = \frac{1-\varepsilon}{\varepsilon}\left[\exp\left(-k\left(t - \frac{\varepsilon L}{u}\right) - \exp(-kt)\right)\right] \qquad (2.35)$$

Goto et al. [32] have derived a shrinking-core leaching model accounting for intraparticle diffusion, external fluid film mass transfer and axial dispersion. Since there is a sharp boundary within a particle between extracted part and nonextracted part, the shrinking-core leaching model may be useful. The model is further simplified to derive an analytical solution. The model was applied to the experimental data for oil extraction from seeds [33, 34]. The following assumptions have been made to derivate the fundamental equations: the solvent flows axially with interstitial velocity, u, through a packed bed in a cylindrical extractor of height, *L*. Pure solvent flows through the bed of particles and the process is considered isothermal. Considering axial dispersion, the material balance on the bulk fluid-phase in the extractor is (Equation 2.36):

$$\frac{\partial C}{\partial t} + u\frac{\partial C}{\partial z} = D_L \frac{\partial^2 C}{\partial z^2} - \frac{1-\varepsilon}{\varepsilon}\frac{3k_f}{R}\left[C - C_i(R)\right] \tag{2.36}$$

Time variation of the solid-phase concentration is equalized with the rate of mass transfer of solute within external film surrounding the particle:

$$\frac{\partial \bar{q}}{\partial t} = \frac{3k_f}{R}\left[C - C_i(R)\right] \tag{2.37}$$

and the diffusion in the outer region is given by:

$$\frac{D_e}{r^2}\frac{\partial}{\partial r}\left(r^2\frac{\partial C_i}{\partial r}\right) = 0 \tag{2.38}$$

Solid phase solute exists within the core, the average value of solid-phase concentration $\bar{q}$ can be given by:

$$\frac{\bar{q}}{q_0} = \left(\frac{r_c}{R}\right)^3 \tag{2.39}$$

Boundary conditions are given as follows. At the core boundary, the concentration in the fluid phase is at its saturation value:

$$C_i = C_{sat} \quad \text{at} \quad r = r_c \tag{2.40}$$

Diffusion flux at the outer surface of the particle is equal to the mass transfer through the external film (Equation 2.41):

$$\left(D_e\frac{\partial C_i}{\partial r}\right)_{r=R} = k_f\left[C - C_i(R)\right] \tag{2.41}$$

Initial conditions are given as follows:

$$r_c = R \quad \text{at} \quad t = 0 \tag{2.42}$$
$$C = 0 \quad \text{at} \quad t = 0 \tag{2.43}$$

Danckwert's boundary conditions at the inlet of the column and the exit condition are given by (Equations 2.44 and 2.45):

$$uC - D_L \frac{\partial C}{\partial z} = 0 \quad at \quad z = 0 \tag{2.44}$$

$$\frac{\partial C}{\partial z} = 0 \quad\quad at \quad z = L \tag{2.45}$$

Kim and Hong [35] have also developed a simplified desorption kinetic model. In this desorption model, a linear irreversible desorption kinetic was assumed, and the dissolution rate of oil components from spearmint leaves was defined by a desorption rate constant. Release of essential oil components from the plant matrix is specified by the desorption rate constant, k. Due to a lack of information about the adsorption isotherm equilibrium, the irreversible linear desorption kinetics was used on the mass balance of oil components in the leaf particles, as presented in Equation 2.46. $\overline{C}$ is the concentration of solute within the pores of leaf particles, W is the weight of leaf particles per extraction, ε is the void fraction of leaf particles, and ρ_p is the apparent density of leaf particles:

$$\frac{d\left(\varepsilon \dfrac{W}{\rho_p} \overline{C} \right)}{dt} = k_e \varepsilon \frac{W}{\rho_p} \overline{C} \tag{2.46}$$

The bulk fluid mass balance for essential oil components in the extractor is written as Equation 2.47. The essential oil components desorbed from the leaf matrixes are instantaneously carried away by the bulk fluid. The bulk fluid with a solute concentration C leaves the extractor with the void volume V at the constant flow rate q.

$$\frac{d(VC)}{dt} = \frac{d\left(\varepsilon \dfrac{W}{\rho_p} \overline{C} \right)}{dt} - qC \tag{2.47}$$

The initial oil concentration in the bulk fluid is zero, and the initial oil concentration inside leaf particles is expressed as $\overline{C}_0$. Therefore, the following boundary conditions are obtained:

$$C = 0 \quad at \quad t = 0 \tag{2.48}$$
$$\overline{C} = \overline{C}_0 \quad at \quad t = 0 \tag{2.49}$$

with

$$\varepsilon \frac{W}{\rho_p} = V_p \qquad (2.50)$$

where V_p is the total pore volume inside leaf particles packed in the extractor. The final solution is obtained as Equation 2.51, which shows the oil concentration change in the bulk fluid over time.

$$C = \overline{C}_0 \frac{V_p}{V} k \left(\frac{e^{-kt}}{\frac{q}{V} - k} + \frac{e^{-\frac{q}{V}t}}{k - \frac{q}{V}} \right) \qquad (2.51)$$

Martínez et al. [22] proposed the Logistic Model (LM) in order to describe the OEC in the extraction of ginger oleoresin with SC-CO$_2$. It is in a fact a modification of the model previously presented by Sovová [23], considering the extract as a mixture of several groups of compounds, classified according to their chemical characteristics.

The model is based on the mass balance of the extraction bed (equations 2.17 and 2.21) and neglects the accumulation and dispersion in the fluid phase because these phenomena have no significant influence on the process when compared to the effect of convection. The logistic equation [36], usually used to model population growth, was chosen to evaluate the variation of the extract compositions. One of its solutions was incorporated in the interfacial mass-transfer term.

The model can then be applied to the solute transfer to the fluid phase and, when assuming the solute represented by a single group of compounds, the OEC (mass of extract, m_{ext}) can be represented by the following equation with only two adjustable parameters (b, t_m):

$$m_{ext}(h = H, t) = \frac{m_t}{exp(bt_m)} \left\{ \frac{1 + exp(bt_m)}{1 + exp[b(t_m - t)]} - 1 \right\} \qquad (2.52)$$

The authors have found that the model was able to describe the OEC very well, where the extracts were considered either as single pseudocomponents or as the sum of groups of compounds. The results were also compared to the ones obtained by applying Sovová´s model and it was observed that the sum of square deviations obtained by using the logistic model was much lower. The model was also successfully applied in the modeling of marigold oleoresin extraction, presenting the best fit to experimental data in comparison with another four different mathematical models [31].

The VTII model [1] integrates the complete mass transfer to the fluid phase side and was previously tested with the extraction of theobromine from cocoa seed shells with SC-CO$_2$, including also analysis for scaling up the process [37]. Additionally, a linear driving force due to the differences of concentrations from the solid to the fluid phase has been postulated.

The model considers the following parameters as sufficient to model the course of one extraction:

- equilibrium distribution between solid and the SCF (adsorption isotherm);
- effective diffusion coefficient or effective transport coefficient defined by the transport model (diffusion in the solid);
- mass transfer from the surface of the solid to the bulk of the fluid phase;
- axial dispersion (effective dispersion coefficient, taking into account inhomogeneities of the fixed bed, the solvent distribution and the influence of gravity).

The equations of the model are given in Equations 2.53 to 2.56. For the fluid phase, the differential mass balance is:

$$\frac{\partial c_F}{\partial t} = D_{ax} \cdot \frac{\partial^2 c_F(z)}{\partial z^2} - \frac{u_z}{\varepsilon} \cdot \frac{\partial c_F(z)}{\partial z} - \frac{1-\varepsilon}{\varepsilon} \cdot \frac{\partial \overline{c}(z)}{\partial t} \tag{2.53}$$

and for the solid phase, the following can be written:

$$\frac{\partial \overline{c}(z)}{\partial t} = a \cdot k_{oG} \cdot \left(c_F(z) - \overline{c}_S(z) \cdot \frac{K(\overline{c})}{\rho_{solid}} \right) \tag{2.54}$$

For the equilibrium between solid and fluid phase:

$$K(\overline{c}_s) = k_1 . \overline{c}_s \, \exp^{-k_2} \tag{2.55}$$

and the overall mass transfer coefficient:

$$\frac{\beta_F}{k_{oG}} = 1 + \frac{Bi \cdot K(\overline{c}_S)}{6} \tag{2.56}$$

Some assumptions have been proposed for the balance in the fluid phase, namely: gradients are neglected in the radial direction, the process is considered isothermal, loading of the SCF is low and a dispersed plug flow regime is achieved. For the solid phase, assumptions are presented as follows: transport can be considered one-dimensional, solid particles are uniform. Real transport phenomena, like pore and solid diffusions, membrane transition, are grouped in one effective transport coefficient (effective diffusion coefficient). Finally, phase equilibrium is assumed at the interface solid/fluid [1].

The equations mentioned for the VTII model were proposed for the extract mixture, neglecting the chemical interactions between the individual components, as well possible synergetic effects between them. In this work, the applicability of the models proposed by Brunner [1] and Martínez et al. [22] have been tested for individual classes of components in the fractionation of rice bran oil.

3. Rice (*Oryza sativa*)

Rice (*Oryza sativa*) is one of the most important cereal crops cultivated worldwide with a production of over 500 million metric tons/year, being also responsible for feeding half the world´s population, especially in the Third World and in developing countries. Rice presents very unique characteristics: brown rice is more caloric than wheat; it presents also the higher digestible energy among all cereals. Rice is also considered the mot versatile crop, being cultivated at higher altitudes and at sea level in the deltas of big rivers in Asia [38].

The genus *Oryza* consists of approx. 20 wild species, but the total number of cultivars is much higher: more than 100,000 rice seed standards are available worldwide. The tropical or indica rice (*Oryza sativa* L.) is being cultivated in southern China for almost seven thousand years and reached Europe in the 8th century. The high number of researches focused on the development of different varieties of rice cultivars demonstrates the efforts of the food policies and governments in order to overcome land constraints, increasing the productivity through higher yields per hectare. Table 3.1 shows the world production in 2003, according to the Food and Agriculture Organization of the United Nations (FAO). More than 90 % of the world production is cultivated in Asia, followed by America and Africa. China and India are responsible for approx. 56 % of the world´s rice production. In Europe, largest producers are Italy (approx. 1.4 Mt/year) and Spain (approx. 0.8 Mt/year) [38].

The structure of a rice grain is presented in Figure 3.1. The most visible part of a rough rice grain is the husk. The caloric value of the husk is rather high and ranges from 3000 to 3500 kcal/kg, making them an important source of energy [39]. However, the most disturbing presence in rice husk is the high proportion of silica, which causes considerable damage to processing equipment through excessive wear of machine parts and interconnecting transfer facilities. When the husk is removed, a thin fibrous layer can be seen. This is called the pericarp, frequently known as the silverskin. This layer is usually translucent or greyish in color. The main function of this layer is to serve as an additional protective layer against molds and quality deterioration through oxidation and enzymes. Sometimes this layer is considered as part of the seed coat but because of its oil content, it is normally considered as the outermost layer of the bran. Immediately under the pericarp is the bran layer or aleurone layer. This part is the main constituent removed in the polishing process. It presents a very low starch content but has a high percentage of oil, protein, vitamins and minerals. Because of its high oil content, the bran is easily affected by oxidation.

Table 3.1. Top 10 rice world producers in 2003 [38].

Country	Rice production (x10^6, metric tonnes)
China	166.00
India	133.51
Indonesia	51.85
Bangladesh	38.06
Vietnam	34.60
Thailand	27.00
Myanmar	21.90
Philippines	13.17
Brazil	10.22
Japan	9.86

The embryo is located at the central bottom portion of the grain, where the grain has been attached to the panicle of the rice plant. This is the living organism of the grain, responsible for the development of a new plant. During milling, the embryo is removed resulting in an indented shape at one end of the milled rice grain. When the husk, the pericarp, the bran and the embryo are removed, what remains is the endosperm (93 wt.-%). It mainly consists of starch with a small concentration of protein and hardly any minerals, vitamins or oil. Because of its high percentage of carbohydrates, its energy value is high [39, 40].

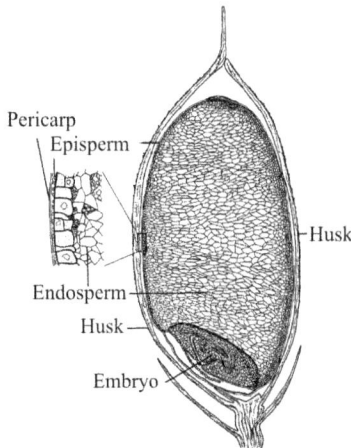

Figure 3.1. Representation of one rice grain (adapted from [39]).

In order to make rice grains susceptible for human consumption, several sequential processes must be carried out, such as cleaning, dehulling, milling and polishing. The dehulled rice is usually denominated as paddy rice, which one is not acceptable for human consumption. By dehulling the paddy rice, the brown rice is obtained. Brown rice can be already used as food, but it presents a high tendency to become rancid due to the lipids present in the embryo and in the grain layers, which ones are very susceptible to oxidize [39, 41-43]. Then, the grain must be polished. Through the polishing of brown rice grains, the embryo and the external layers are separated and the final product, the polished rice, is obtained. The residual material of the polishing process is collected and denominated rice bran [39, 40].

Rice bran is considered one excellent source of vitamins, free fatty acids (FFA), iron, magnesium, proteins, lipids and phosphorus, being considered an extremely popular source of dietary fiber because of the hypocholesterolemic properties of its oil fraction, especially due to the presence of sterols and their derivatives, especially γ-oryzanol (see further discussions in Chapter 3.1.1).

In Brazil and other developing countries, rice bran is a product of low commercial value, used normally as an ingredient of animal ration, as an organic fertilizer and very rarely for oil extraction. Another use of rice bran is in a "multi-mixture", a toasted flour (generally in home-made form) consisting of food industry wastes: as a part of one Brazil's social program, the mixture is distributed to children of low family income. More recently, rice bran has been used in normal human diet (bakery and desserts' formulations), but the industry has not yet completely adapted itself to rice bran production for human consumption. Prior to its use, the bran must be very quickly stabilized, due to the presence of active lipases and peroxidases, which ones are responsible for its rapid rancidity, i. e., the increase of the total FFA content [44]. Depending on the characteristics of the bran, within two days after processing, RBO extraction is no longer economically feasible, and after three or four days, it is no longer possible to use the raw bran as animal feed (increase in the acidity up to 70 wt.-%) [45].

Rice bran is usually thermally stabilized: the most used classic methods include dry heat, moist heat and moist heat on press stabilization [44, 46]. Heated rice bran at 120 °C for 15 min can present up to 13 % residual lipase activity. Thus, McPeak [45] developed a continuous rice bran extruder/pelletizer apparatus with appropriate heating in order to stabilize the bran by applying minimal temperatures of 130 °C.

The work of Lakkakula et al. [46] presented the use of ohmic heating in order to stabilize rice bran, enhancing consequently extraction yields of RBO. Ohmically heated/stabilized rice bran samples yielded 49–92 wt.-% of total lipids through liquid extraction as compared to 53 wt.-% of total lipids with untreated samples. When the electrical field strength was increased from 60 to 100 V/cm, the extraction yield for all moisture contents increased, especially with raw bran. The temperature effect was also evaluated, what is important for processes involving extraction of termolabile components.

Other available methods include the stabilization by microwaves and by using chemical products, such as acetic and hydrochloric acids, propanal and acrylonitrile [47-49].

3.1. Rice bran oil (RBO)

Once the rice bran is stabilized, RBO can be extracted. As other oilseeds, the oil is usually extracted with hexane (see Chapter 3.2).

According to official data [38], RBO commercial production through the extraction with hexane was about 783,000 ton in year 2000. In year 2002, India was the first RBO producer worldwide, achieving up to 500,000 ton crude oil and more than 400,000 ton refined oil [50]. However, RBO world production is approximately only 1 wt.-% of the total annual produced vegetable oils, due mainly to the difficulties in the processing of rice bran, as discussed in the previous section. Other potential producers are Japan, China, Korea and Thailand. The oil is considered a valuable cooking oil, being extensively consumed in Asian countries, including Japan (approx. 80,000 ton annually), China, Korea, Thailand, India, Taiwan and Pakistan [51, 52]. However, its consumption is still negligible in Western countries.

RBO presents a better frying performance in comparison with partially hydrogenated soybean oil (HSBO), which is still the most used edible oil worldwide. The study was conducted at Riceland's Research & Technical Center [53] and simulated restaurant cooking conditions, where the oils were submitted to continuous frying. Their frying performances were evaluated through a Food Oil Sensor (FOS), designed to verify the dielectric constants of fats. As oil undergoes oxidative and thermal breakdown, leading to a consequent formation of FFA, the dielectric constant increases up to maximal level accepted, which is 5. Then, the oil must be replaced. Results have shown that FOS values for HSBO steadily increased during the study, reaching level 5 after approx. 168 frying hours, while RBO reached around 3 in the same frying time. FFA levels increased up to 8 wt.-% after the same time for HSBO. For RBO, the amount achieved reached 3 wt.-% in the same conditions. Additionally, HSBO started smoking during the experiments. Blind sensorial tests were also performed and it was observed that french fries fried with RBO were crispier and much less greasy than the ones fried using HSBO. Off-flavors were also detected after the use of soybean oil.

Chemically, RBO is one of the most nutritious and healthful vegetable oils available. The abundance of natural bioactive components, such as fats, oryzanols, tocols and sterols, with a relatively high FFA composition (Table 3.2), makes it a very attractive source of nutraceutical components. Recently, it has been found that RBO is as good or even better than other vegetable oils in reducing serum cholesterol and risk of heart diseases, due mainly to the synergetic effects involving all its components. Their functional properties are extensively discussed in the literature and will be summarized in Chapter 3.1.1.

Table 3.2. Chemical composition of crude RBO [54].

Components	Composition
TG, (HPLC area, %)	68.0 – 72.0
FFA, (HPLC area, %)	5.0 – 10.0
Oryzanols, (wt.-%)	1.0 – 2.0
Free sterols, (wt.-%)	0.3 – 0.5
Sterol fatty acid esters, (wt.-%)	3.3 – 3.9

One interesting and very promising application of RBO is the production of biodiesel, composed mainly by fatty acid methyl esters (FAME). In the USA, soybean oil is the most commonly used biodiesel feedstock, whereas rapeseed (canola) and palm oil are the most common sources in Europe and in tropical countries, respectively.

Zullaikah et al. [55] studied a two-step acid-catalyzed methanolysis process for the conversion of RBO oil into FAMEs. The first step was carried out at 60 °C and, depending on the initial FFA content of oil, 55–90 % FAME content in the reaction product could be obtained. The organic phase of the first step reaction product was then used as raw material for the second acid-catalyzed methanolysis (at 100 °C). In the end, more than 98 % FAME in the product were obtained in less than 8 hours. Distillation of reaction product resulted in 99.8 % FAME (biodiesel), with recovery of more than 96 %. The residue contained enriched nutraceuticals, such as oryzanol (16–18 %), mixture of phytosterols, tocols and steryl esters (19–21 %), which ones can be successfully purified, in order to aggregate higher values to these residual fractions.

RBO waste residues have shown a high potential to be used in the production of vanillin. A new technology of transforming ferulic acids into vanillin was developed by using two different fungal straints (*Aspergillus niger* and *Pycnoporus cinnabarinus*) [56]. Ferulic acids are very common and abundant in plant cell walls, occurring thus in common agricultural waste residues, such as cereal brans, sugar beet pulp and some edible oils, including RBO. The maximal vanillin concentrations reached 2.8 g/l, with a molar yield of approximately 62% after 72 h conversion time. Through ^{13}C isotope analysis, the authenticity of the vanillin product was confirmed.

3.1.1. Key-Components

Four different classes of compounds were chosen for the modeling and separation analysis of RBO when employing SC-CO_2 as a solvent: FFA, TG, oryzanols and sterols. A brief overview on these classes of compounds is presented in the following sections.

3.1.1.1. Fatty acids

In chemistry, especially biochemistry, a fatty acid is a carboxylic acid often with a long unbranched aliphatic chain, which can be either saturated or unsaturated. Fatty acids are mostly present in fats and vegetable oils; they can be bound or attached to other molecules, such as in TG or phospholipids. Several hundreds forms have been identified, but the chain lengths range from 2 to 80 – commonly from 12 up to 24 carbon atoms. Their well known linear structure is presented in Figure 3.2.

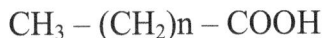

$$CH_3 - (CH_2)n - COOH$$

Figure 3.2. Linear structure of fatty acids.

Industrially, fatty acids are produced by the hydrolysis of the ester linkages in a fat or vegetable oil, with additional removal of glycerol molecules. When they are not attached to other molecules, they are commonly defined as free-fatty acids (FFA). Fatty acids and their derivatives provide one of the essential materials needed to the preparation of several foodstuff, cosmetics and pharmaceuticals [57]. Fatty acid compositions depend on the oil sources. Table 3.3 shows a comparison between fatty acid compositions of different vegetable oils.

Table 3.3. Fatty acid composition of different vegetable oils.

Oil	Composition (wt.-%)						
	Myristic ($C_{14:0}$)	Palmitic ($C_{16:0}$)	Stearic ($C_{18:0}$)	Oleic ($C_{18:1}$)	Linoleic ($C_{18:2}$)	Linolenic ($C_{18:3}$)	Others
Palm oil[a]	1.0	46.2	5.9	34.8	11.8	0.1	0.2
Crude RBO	tr[b]	19.9[b]	1.8[b]	40.1[b]	36.3[b]	1.9[b]	-
	0.2[c]	25.0[c]	1.5[c]	35.0[c]	26.0[c]	1.7[c]	10.6[c]
Refined RBO[b]	0.9	18.2	1.5	38.5	35.6	2.7	2.6
Rapeseed oil	tr	4.0	2.0	62.0	22.0	10.0	-
Soybean oil	tr	11.0	4.0	24.0	54.0	7.0	-

[a] - obtained from Loncin [57]; [b] - obtained from Rodrigues et al.[58]; [c] - obtained from García et al.[59].

Some fatty acids can not be produced by the human body, like linolenic and alpha-linolenic (ω-3) acids. In this case, substrates must be supplied in food, such as vegetable and fish oils, and they are called essential fatty acids. After absorption, essential fatty acids are primarily used to produce hormone-like substances, in order to regulate a wide range of functions. They are important in several human body systems, including the immune system and in blood pressure regulation, since they are used as precursors for the different classes of compounds, such as prostaglandins. Numerous studies in humans and animals have demonstrated that oils containing saturated fatty acids raise serum total cholesterol and, in particular, low density lipoprotein cholesterol (LDL-C) levels, while those enriched in unsaturated fatty acids lower LDL-C levels when replacing saturated fat [60]. Studies performed approximately 40 years ago have demonstrated that the fatty acid components and cholesterol in the diet are the primary determinants of diet induced hypo- or hypercholesterolemia. Nowadays, it has been explained that the hypolipidemic effect of RBO is not entirely explained only by its fatty acid composition, but for the synergy between each individual class of components. In fact, several studies have also indicated hypercholesterolemic properties of some unsaponifiables, especially plant sterol components [40, 60-62].

3.1.1.2. Triglycerides (TG)

TG are the predominant component of most food fats and oils, formed through the esterification of one glycerol and three fatty acids molecules, being palmitic, oleic and linoleic acids the most common ones. Chemically, fats and oils are either simple or mixed glyceryl esters of organic acids belonging to the fatty-acid series, derived from both plant and animal sources [40, 47]. Two different molecules of TG are represented in Figure 3.3.

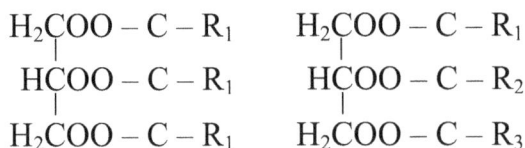

$$
\begin{array}{ll}
H_2COO - C - R_1 & H_2COO - C - R_1 \\
HCOO - C - R_1 & HCOO - C - R_2 \\
H_2COO - C - R_1 & H_2COO - C - R_3
\end{array}
$$

Figure 3.3. Chemical structures of TG (R_1, R_2 and R_3 represent different FFA).

TG are a group of natural organic substances that form an important part of the human diet, playing an important role in metabolism as energy sources. They contain more than twice as much energy as carbohydrates and proteins. In the intestines, TG are split into glycerol and fatty acids (lipolysis) by lipases and bile secretions, which can then be absorved.

TG correspond to up to 95 wt.-% of fat and oils compositions and are also, together with their fatty acid esters, important raw materials in the pharmaceutical, food and chemical industries, where they are used for emulsification, starch complexing, aeration, as defoaming agents and as oil stabilizers. One important application is the production of biodiesel: TG is the main raw material for the its manufacturing, being submitted to transesterification reactions in order to obtain the desired fatty acid methyl/ethyl esters [41, 63-66].

3.1.1.3. Sterols

Sterols, also referred as steroid alcohols, are a class of substances that contain the common steroid nucleus plus an 8 to 10 carbon side chain and an alcohol group and are an important component of the membranes of multicelular organisms. Although sterols are found in both animal fats and vegetable oils, there is a substantial difference biologically between those occurring in animal fats and those present in vegetable oils [67-69]. Cholesterol is the primary animal fat sterol and is only found in vegetable oils in trace amounts. Vegetable oil sterols are collectively termed phytosterols. Sitosterol, campesterol and stigmasterol (Figure 3.4) are the best-known vegetable oil sterols.

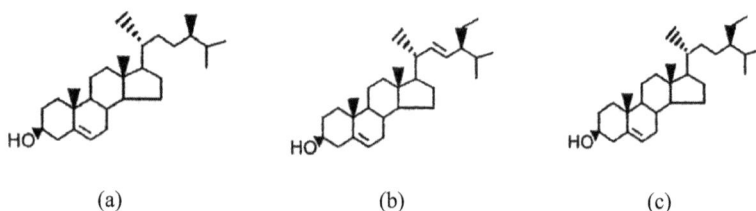

(a)   (b)   (c)

Figure 3.4. Structures of campesterol (a), stigmasterol (b) and sitosterol (c).

Sterols are minor components of all vegetable oils and they correspond to the major fraction of the unsaponifiable fraction of several oils, such as corn fiber, palm, soybean, wheat bran and rice bran oils [54, 67]. According to Orthoefer [62], the RBO unsaponifiable fraction is composed by 43 % phytosterols, 10 % steryl esters and 1 % tocopherol (wt.-%).

Since free sterols are not soluble in food systems, they are usually esterified with fatty acids to obtain sterol-esters, which present high solubility in fats, allowing them to be added to several food systems as functional ingredients. Phytosterols were approved as GRAS ("Generally Recognized As Safe") by the US Food and Drug Administration (FDA) for the use in margarines in 1999 [54, 70]. Since then, the enrichment of foods with these nutraceutical compounds is highly recommended due to the increased consumer demand for such nutraceutical foods. Their cholesterol-lowering properties are well known and were first shown in the early 1950s. Years later, numerous studies have still reported the hypocholesteremic potency of plant sterols [41, 62, 67, 69, 71, 72].

3.1.1.4. Oryzanols

Oryzanol was identified in 1954 [73]. As it was first isolated from RBO (genus *Oryza*) and contained one hydroxyl group, was conveniently named oryzanol. In the early years, it was presumed that oryzanol was only a single component. Later it was found through chromatographic and mass-spectrometric methods that oryzanol is in fact a mixture containing ferulic acid esters of unsaturated triterpenoid alcohols and plant sterols [67, 74], as presented in Figure 3.5.

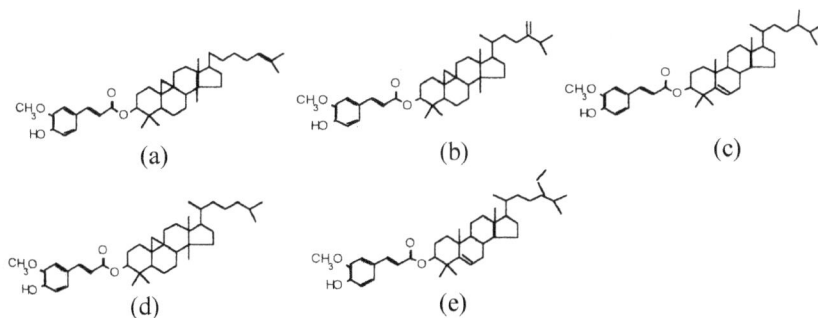

Figure 3.5. Structures of the ferulic acid esters known as γ-oryzanols (a=cycloartenyl ferulate; b= 2,4 methylene cycloartenyl ferulate; c= campesteryl ferulate; d= cycloartanyl ferulate; e=β-sitosteryl ferulate).

Oryzanol components were also isolated from corn, wheat and linseed grains [75], but RBO is the most accessible natural source of oryzanols and, although their concentrations may vary according to the origin of the bran, they can achieve up to 2.0 wt.-% of the oil, the highest concentrations between the mentioned cereals. Cycloartenyl ferulate, 2,4 methylene cycloartenyl ferulate and campesteryl ferulate are the three major components and account for approximately 80 wt.-% of γ-oryzanol in RBO [74]. The isolation of oryzanol from RBO soapstock has been also investigated, especially in rice-producing countries, like China, Japan, India, Thailand and the United States. It is possible to find more than 40 filled patents on the isolation of oryzanol from RBO soapstock worldwide [74].

Physiological effects that have been shown to be associated with oryzanol contents are decreasing plasma cholesterol, platelet aggregation, hepatic cholesterol biosynthesis, cholesterol absorption and aortic fatty streaks formation. Oryzanol has also been used in order to increase fecal bile acid excretion and in the treatment of nerve imbalances and disorders of menopause [76, 77]. In Japan, oryzanol is used as an antioxidant in the conservation of oils and beverages, presenting a synergetic effect in association with vitamin E [40, 78]. It is also described as a multifunctional cosmetic agent: its properties consist on the stimulation of sebaceous glands and in the adsorption of UV radiation, especially UV-A (320-340 nm) and UV-B rays (280-320 nm) [79, 80]. Therefore, oryzanols are employed in the formulation of sunscreens and hair care products, which ones are already available in the market [74].

3.2. Edible oils´ extraction and processing

Commercial sources of edible oils and fats include oilseeds, fruit pulps, fish and other animal species. According to official data [81], the major oilseeds world production in 2002 was approx. 326 million metric tons.

The great interest on the consumption of vegetable oils and animal fats is not only related to their high concentrations of triacylglycerols; they also contain several groups of other substances, which may be useful as nutrients, the nutraceutical components. At the same time, other minor substances can be considered undesired, affecting especially sensorial (such as taste, color and odor) or even functional properties [81, 82]. The most common groups of these minor components are presented in Table 3.4.

Table 3.4. Groups of minor substances present in crude fats and oils.

Substances	Typical examples	Deteriorating effect
Oxidation products	Volatile aldehydes, ketones, waxes, hydrocarbons	Off-flavors
FFA	Saturated and unsaturated fatty acids	Lower stability, impaired functional properties
Phospholipids	-	Lower oxidative stability
Pigments	Carotenoids, globulins, chlorophylls	Affect sensory characteristics
Metal salts	Iron and cupper compounds	Affect oxidative stability

The method chosen for edible oil extraction depends usually on the natural aspects of the raw material and on the capacity of the industrial plant. The most widely employed extraction methods for several varieties of oilseeds are pressing followed by organic solvent extraction, which contribute to almost half of the total production of vegetable oils around the world. The crude oil extracted is a complex mixture, formed especially by sterols, tocopherols, FFA, mono-, di- and triglycerides, phosphatides, pigments, as well as traces of glycolipids, metals, flavonoids, tannins, among others [81, 82].

Commercial-grade hexane, a paraffinic petroleum fraction, has been the solvent of choice for organic solvent extraction of edible oils due to both practical and economical reasons: it presents a fairly narrow boiling point (63–69 °C) and is an excellent oil solvent, being also easily recovered. However, such desirable properties of hexane are also directly responsible for serious problems, since it is pollutant, toxic, flammable and explosive. Hexane, when inhaled by humans, is capable of dissolving in neural lipids, affecting then the nervous system [83]. *n*-Hexane, the main component of commercial hexane, is listed as No.1 of 189 hazardous air pollutants (HAPs) by the US Environmental Protection Agency. Edible oil processing industries are considered a major source, since they are responsible for more than 10 ton per year of HAPs, requiring then a Federal Operating Permit in the United States [83]. Even when considering a moderate loss of 0.15 % per ton of feed, an industrial plant with an average daily processing capacity of about 75 ton would result in 42 ton of hexane emissions in the environment every year. All these concerns have stimulated the interest in the research on alternatives to hexane as an extraction solvent [81].

Researches on alternative solvents and methods for the extraction of edible oils had focused mainly on alcohols, halogenated hydrocarbons, hydrocarbons, water and supercritical fluid extraction, which one will be discussed in details in Chapter 3.3.

Monsoor et al. [84] and Monsoor and Proctor [85] have reported aqueous extraction procedures for RBO. Phenolic content of extracts at different temperatures (20 to 60 °C) were high (63.28 to 82.51 mg/100 ml extract) and their antioxidant capacity was similar to the ones obtained from servings of fruits and vegetables.

Proctor and Bowen [86] have compared hexane and isopropanol extractions of RBO at ambient temperatures. They have concluded that isopropanol was as effective as hexane in the extraction of RBO and FFA levels were very similar, even when up-scaling the process (from 2 to 30 g of rice bran). Besides, oil extracted with isopropanol was significantly more stable to heat-induced oxidation that the hexane extracts, probably due to the fact that antioxidants, like tocopherols, were more easily extracted by isopropanol.

The application of limonene as a solvent for RBO extraction was investigated by Mamidipally and Liu [83]. RBO extraction was also carried out with hexane in order to compare the efficiency of both solvents. Results have shown that limonene extracted a significantly higher amount of oil than hexane under any given condition and the quality oil characteristics were very similar. The authors mentioned that the oxidation stability of limonene was about to be tested. As a disadvantage, limonene is more expensive than hexane. However, the authors concluded that, taking into account the costs associated with environmental policies, RBO extraction with limonene may become a very attractive and competitive process.

In order to close the cycle, the remaining defatted rice bran can be used for the production of polylactic acid, one of the most promising biodegradable plastics. Since defatted rice bran contains polysaccharides as starch and cellulose, saccharification with amylase and cellulase to lactic acid fermentation were coupled and investigated by Tanaka et al. [87]. The yield based on the amount of soluble sugars after 36 hours hydrolysis was 78 %, reaching an optic purity of 95 % for the produced D-lactic acid. Although the results obtained were very promising, further studies are still necessary for the cost analysis, especially regarding to enzyme costs.

The removal of nontriglyceride fatty compounds is normally defined as refining, although in the USA this denomination is applied to pre-treatment and deacidification or neutralization operations. In many other countries it represents a complete series of treatments, including also bleaching and deodorization [81]. Chemical and physical are the most used methods for refining edible oils. A simplified scheme of oil processing can be seen in Figure 3.6.

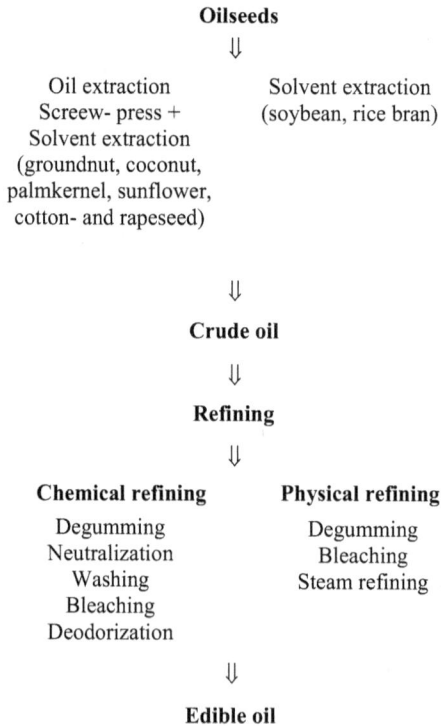

Oilseeds

⇓

Oil extraction	Solvent extraction
Screew- press +	(soybean, rice bran)
Solvent extraction	
(groundnut, coconut,	
palmkernel, sunflower,	
cotton- and rapeseed)	

⇓

Crude oil

⇓

Refining

⇓

Chemical refining	**Physical refining**
Degumming	Degumming
Neutralization	Bleaching
Washing	Steam refining
Bleaching	
Deodorization	

⇓

Edible oil

Figure 3.6. Simplified flowsheet of conventional oil processing (extraction and refining).

Crude fats and vegetable oils consist mainly of TG, along with FFA, which are virtually absent in oils/fats of living tissues. However, they can be formed through enzymatic action (lipase) after the animal has been slaughtered or after harvesting the oilseed. As previously described, the release of short-chain fatty acids by hydrolysis is responsible for the development of an undesirable rancid flavor [81, 82]. It is also well known that FFA are much more susceptible to oxidation than their esters. This lipid oxidation leads to an oxidative rancidity in edible oils and fat containing foods and, in order to preserve the quality of the oil, any increase in its acidity must be avoided.

The deacidification process presents the highest economic impact in the oil production and its efficiency is determinant for the subsequent refining steps. Removal of FFA from crude oils is considered a very delicate and difficult stage in the whole refining cycle, because it is responsible for the final product's quality [81]. Chemical, physical and miscella deacidification methods are usually employed industrially for deacidification of edible oils. These conventional methods are well documented and discussed elsewhere [81, 82, 85, 88, 89] and some of their features and limitations are summarized in Table 3.5.

Table 3.5. Industrial methods for deacidification of edible oils (adapted from [81]).

Features	Limitations
Chemical deacidification	
Versatile – producing acceptable quality oil Multiple effects – purifying, degumming, neutralizing and partially decolourizing the oils	Excessive loss of neutral oil (high FFA content), soapstock – low commercial value, neutral oil losses due to hydrolysis
Physical deacidification	
Suitable for high FFA contents, low capital and operating costs, higher oil yields, elimination of soapstock and reduced effluent streams, improved quality (FFA)	Pretreatments are very stringent, not suitable for heat sensitive oils, controlled rate of FFA removal, chances of thermal polymerization
Miscella deacidification	
Lower strength of caustic solution, increased efficiency of separation, minimum oil occlusion in soapstock, washing water eliminated, superior color of final product	Higher investment – cost intensive, solvent losses (careful operation and constant maintenance), more suitable for integrated plant and refining plant

During the chemical deacidification process, considerable losses of neutral oils, vitamins and tocopherols can be achieved. Besides, the disposal and utilization of high amounts of resulting soapstock may cause serious problems of waste management [81]. On the other hand, physical deacidification can lead to good results only when using good quality oils as starting material. The application of miscella deacidification is also limited by the two-stage solvent removal system. According to Anderson [90], these methods are not recommended for oils containing more than 8-10 wt.-% FFA. RBO of high FFA content, for instance, can present neutral losses up to or higher than 50 % during chemical deacidification [81, 82]. In order to overcome these drawbacks, alternative approaches are needed, namely: biological deacidification, reesterification, solvent extraction, membrane processing and SFE, which one will be extensively discussed during the present work. Main features and drawbacks of the first four techniques listed are summarized in Table 3.6.

Table 3.6. Alternative methods for deacidification of edible oils.

Features	Limitations
Biological deacidification Employing whole cell microorganisms (selectively assimilation of FFA)	Linoleic acid and short-chain FA not used, inhibition of microbial growth, fatty acids use depends on its solubility in water
Biological deacidification Enzyme reesterification – Lipase reesterification Increased oil yield, low-energy consumption, mild operational conditions	High cost (enzymes)
Reesterification With or without the aid of catalyst, suitable for high FFA-oils, increased oil yields	Random reesterification, thermal polymerization, costs
Solvent deacidification Extraction at ambient temperature and atmospheric pressure, easy separation separation	Higher capital cost, energy intensive operation, incomplete deacidification (TGs solubility increase with FFA in feed)
Membrane deacidification Low-energy consumption, ambient temperature operation, no addition of chemicals, retention of nutrients and other desirable components	Molecular weight difference is small, non-availability of suitable membrane with high selectivity, low permeate flux

3.3. SFE applications to RBO – State of the art

Critical fluids processing can be perfectly used in several modes for producing/isolating nutraceutical ingredients from edible oils. In the field of SFE, various researchers proposed the use of SC-CO_2 for RBO extraction, as well as for the refining/deacidification process. These studies ranged from simple batch extractions (carried out usually by using lab-scale equipments) to RBO fractionation using countercurrent columns. An overview on published works dealing with extraction and fractionation of RBO is presented below.

3.3.1. Extraction

Most studies of the applicability of SFE in RBO extraction have focused on the yield of extractable material (mainly lipid components). The supercritical fluid pressure, temperature, solvent flow rate, i. e., CO_2 consumption, and time of operation are optimized in order to achieve higher extraction yields with lower solvent and energy demands. The evaluation of the best conditions to be applied to one determined system is generally performed through the analysis of the overall extraction curves obtained.

García et al. [59] performed the separation of waxes and long chain fatty acids from rice bran at 28 MPa, with the temperature varying from 40 to 70 °C. Highest extraction yields were obtained at the highest operational conditions investigated (28 MPa and 70 °C). Compared to extracts obtained with hexane, the supercritical extract was lighter in color and richer in waxy components, what can be observed in Table 3.7.

Table 3.7. Comparison between hexane and SC-CO_2 extracts [59].

Solvent	Oil (g/100g extract)	Waxes (g/100g extract)
Hexane (70 °C)	86	14
SC-CO_2 (28MPa-70 °C)	64	36

Concerning to fatty acid and alcohol compositions, the same components were identified in both extracts. However, they were found at different ratios: hexane was responsible for a higher dissolution of palmitic, oleic, linoleic and linolenic acids (see Table 3.8), while the fraction soluble in CO_2 was composed mainly by long-chain fatty acids (C_{20}-C_{34}), such as tetracosanoic and docosanoic acids. The poor dissolution of smaller molecular weight fatty acids (C_{14}-C_{18}) in carbon dioxide, compared to their dissolution in hexane, was explained by the authors considering that the fatty acids solubilities in SC-CO_2 are lower than in hexane at the experimental conditions investigated. According to this premise, CO_2 should have also extracted less long-chain fatty acids (C_{20}-C_{26}, not shown in Table 3.8) than hexane. However, experimental results have shown the opposite. The authors suggested that could be possibly due to the existence of weak interactions between these components (C_{20}-C_{26}) and the solid matrix, which favored their extraction with carbon dioxide.

Table 3.8. Fatty acid composition of hexane and SC-CO_2 extracts [59].

Fatty acids	SC-CO_2 (wt.-%)	Hexane (wt.-%)
Myristic ($C_{14:0}$)	0.4	0.2
Palmitic ($C_{16:0}$)	17.0	25.0
Stearic ($C_{18:0}$)	1.1	1.5
Oleic ($C_{18:1}$)	19.3	35.0
Linoleic ($C_{18:2}$)	16.6	26.0
Linolenic ($C_{18:3}$)	0.2	1.7
Others	45.4	10.6

The purpose of the work published by Shen et al. [91] was to measure the effects of temperature, pressure, and flow rate of SC-CO_2 in the extraction and fractionation of RBO. Column beds (300 g) of rice bran were submitted to SFE with CO_2 at a solvent flow rate of 2.5 kg/h, temperatures of 0-60 °C and pressures of 17-31 MPa, reaching up to 6 h process. The extracted total oil, the FFA, tocopherol, sterols (campesterol, stigmasterol, sitosterol), and oryzanol components were measured at different time intervals. Figure 3.7 presents the effects of temperature and pressure on oil yield as a function of the amount of CO_2 used. For comparison purposes, oil yield is reported as a percentage of the amount extractable by hexane. Oil was extracted linearly up to 80 wt.-% of the hexane-extractable amount at 24 MPa (20 °C and 40 °C) and 31 MPa (40 °C), or remained linear throughout the whole extraction experiment at 24 MPa (0 °C and 60 °C) and 17 MPa (40 °C). The extractions at 31 MPa and 40 °C provided the highest oil yield, achieving approx. 97 wt.-% of hexane extractable oil.

The authors described the extraction of oil components through analysing apparent partition coefficients between oil and solvent phases. The observed differences in partition coefficients provided a basis for refining and fractionation of RBO: at 17 MPa and 40 °C, the partition coefficient of FFA was 3.8 times that of TG, 3.2 times that of tocopherol and 11.5 times that of oryzanol, which were the largest differences observed among all extraction conditions investigated. These calculations were similar to trends observed by Maheshwari et al. [92], who showed that the lower the density of CO_2, the more efficiently FFA are separated from TG.

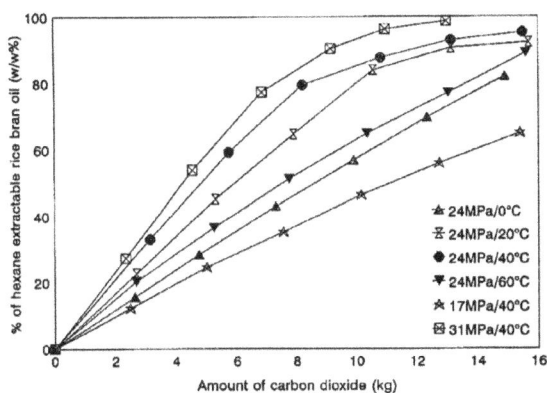

Figure 3.7. SC-CO_2 extraction of RBO at different pressures and temperatures with a solvent flow rate of 2.5 kg/h [91].

One year later, Shen et al. [93] published a second paper on the extraction and fractionation of RBO. The aim was to continuously produce an oil of enhanced composition using a second-stage expansion column after primary SC-CO_2 extraction and to utilize the data from the expansion column to evaluate the solubility of the oil, partition coefficients, and the selectivities of its components, as a function of temperature, pressure, and density under the investigated conditions. In the first stage of the investigated two-stage process, crude RBO was extracted with SC-CO_2 from 300 g of rice bran. Oil-laden CO_2 from the extractor (24.1 MPa and 40 °C) flowed continuously to a second-stage column, where an oil phase (raffinate) was separated from the solvent at various controlled temperatures and pressures. Measurement of the compositions of raffinates and extracts allowed calculation of partition coefficients of TG, FFA, tocopherol, sterols, and oryzanol. Oryzanol presented a partition coefficient that was lower than or equal to the partition coefficient of TG under all CO_2

conditions at 40 °C. At CO_2 densities >0.6 g/ml it was possible to separate RBO into a high oryzanol fraction and a low oryzanol fraction. The high oryzanol fraction presented also a reduced FFA content (up to 50 wt.-%).

Kuk and Dowd [94] studied the extraction of rice bran lipids with $SC-CO_2$. Different experimental conditions were employed in order to investigate pressure effects on the extraction yields (48-62 MPa and 70-100 °C), which were in the range of 20 wt.-%. Temperature effects were shown: lipid extraction yield increased with an increase in temperature under isobaric conditions. Through the analysis of crude RBO samples, the authors have found that most of the crude oil from either hexane or $SC-CO_2$ extraction was composed by glycerides (from 70 to 80 wt.-%), FFA up to 5 %, sterols from 1 to 3 %, trace amounts of glycol- and phospholipids, and the rest was composed by unidentified wax. The wax components were assumed to be composed by saturated fatty alcohols and alkanes (C_{24}-C_{32}).

Kim et al. [95] studied the extraction of essential fatty acids from domestic brown rice bran and the extracts were analyzed through gas chromatography-mass spectrometry techniques. As expected, the extracted amount of RBO was dependent upon the operating pressure and temperature, and the fatty acid composition varied with the reduced density of the $SC-CO_2$. For comparison purposes, organic solvent extraction was also performed. The solvent chosen was composed by a mixture chloroform:methanol:water (1:2:0.8, volume ratio). The maximum amount of unsaturated fatty acids extracted was 83.4 wt.-% at 40 °C and 27.6 MPa, the optimum operating condition for the separation of essential fatty acids. About 80 wt.-% polyunsaturated fatty acids (PUFA) was extracted in SFE, whereas about 60 wt.-% PUFA was extracted with the mentioned organic solvent mixture, which proved that SFE was more suitable to fatty acids extraction.

Comparison of supercritical fluid and solvent extraction methods in extracting oryzanols from rice bran have been reported in the literature by Xu and Godber [96]. A solvent mixture with 50 % hexane and 50 % isopropanol (vol/vol) at 60 °C for 45-60 min produced the highest yield of oryzanol (1.68 mg/g of rice bran) among the organic solvents tested. SFE experiments conducted during 25 minutes at 68 MPa and 50 °C presented the highest oryzanol yields (5.39 mg/g of rice bran), approximately 4 times higher than the ones obtained through organic solvents extraction.

Perretti et al. [97] have evaluated the use of SFE technology for the recovery of rice by-products and novel conversion processes to manufacture value-added food products from agricultural wastes. The investigated raw materials were cleaned rice, husks, rice grits, green rice, white rice, rice bran, chalky rice, broken and discolored rice grains. Figure 3.8 shows a flowsheet of the rice milling process investigated, showing also the origin of each raw material.

Conditions were studied to extract oil from the mentioned products and byproducts in order to increase the concentration of antioxidants (especially oryzanols) in oil. The authors achieved the highest amount of oryzanols (18 mg/g raw material) through SFE of RBO at 68 MPa and 80 °C. Other rice by-products presented also high oryzanol contents, like green rice (17.7 mg/g sample), discolored grains (11.3 mg/g sample) and broken rice (8.5 mg/g sample).

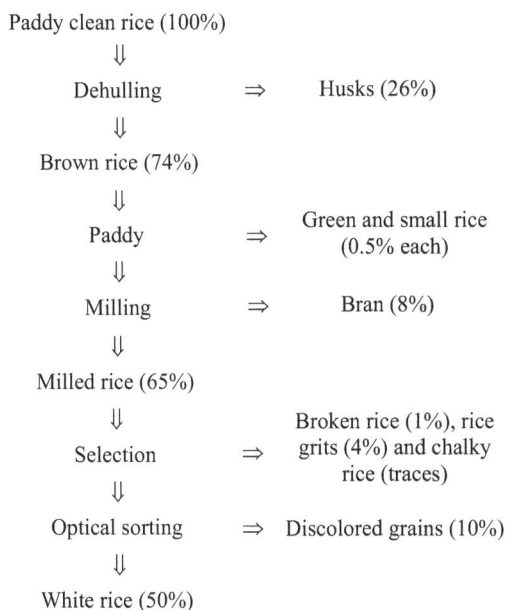

Paddy clean rice (100%)
⇓
Dehulling ⇒ Husks (26%)
⇓
Brown rice (74%)
⇓
Paddy ⇒ Green and small rice (0.5% each)
⇓
Milling ⇒ Bran (8%)
⇓
Milled rice (65%)
⇓
Selection ⇒ Broken rice (1%), rice grits (4%) and chalky rice (traces)
⇓
Optical sorting ⇒ Discolored grains (10%)
⇓
White rice (50%)

Figure 3.8. Rice milling process: scheme and origin of each sample [97].

The recent work of Sparks et al. [98] dealt with the extraction of rice bran lipids and the experiments were performed with SC-CO_2 and liquid propane. In order to provide a basis for extraction efficiencies, organic accelerated solvent extractions with hexane were performed at 100 °C and 10.3 MPa. Parameters evaluated in the experiments were 45, 65 and 85 °C and from 20 to 35 MPa. For CO_2 experiments, extraction efficiencies were proportional to pressure and inversely proportional to temperature. Maximal yield obtained through CO_2 extraction was 22.2 wt.-% at 45 °C and 35 MPa. When the extraction was performed with propane, the maximal yield obtained was 22.4 wt.-% at 0.76 MPa and ambient temperature. Maximal extraction yields from both high pressure processes were compared to hexane extraction, which achieved 26.1 wt.-% yield. They have also performed an economical analysis on the possibility of using SC-CO$_2$ and propane in the extraction of RBO in Mississippi, USA. The authors concluded that none of the evaluated extraction techniques was economically feasible for the treatment of 438.9 kg (propane and SC-CO$_2$) and 408 kg of rice bran (hexane) per day. Capital investments (only considering the extraction equipments) were US$190,000; US$225,000 and US$220,000 for propane, SC-CO$_2$ and hexane extraction processes, respectively. As a final consideration, they concluded that by increasing capacity, "economies of scale" may have a positive effect on lowering the units´ costs, but no further analysis was presented.

3.3.2. Countercurrent column extraction

Examinations on the potential of a continuous countercurrent fractionation process employing SC-CO$_2$ for the refining and deacidification of RBO have been performed and are already available in the literature. Mostly, these works investigated the removal of FFA in the extract stream, while the most valuable lipid components (TG, sterols and oryzanols) were enriched in the raffinate fractions [54, 70, 99-101], minimizing then the losses typical in the chemical deacidification process.

The crude RBO deacidification utilizing SC-CO$_2$ as solvent was firstly performed by Dunford and King [54] by using a 1.64 m height and 260 cm^3 internal volume packed column. It is important to mention that no extract reflux was used. The feed composition profile can be seen in Table 3.2. The authors employed different pressure (20.5 to 32.0 MPa) and temperature (45 to 80 °C) levels for the isothermal operation of the column in order to investigate the composition profiles of the resultant fractions. Low-pressure and high-temperature conditions (20.5 MPa and 80 °C – CO$_2$ density 0.6 g/ml) proved to be favorable for minimizing TG and phytosterols concentrations in the extract samples during the FFA removal from crude RBO. At these conditions, FFA concentration achieved 37 wt.-% in the extract fraction. This could be explained by the higher solubility and selectivity for FFA in high-pressurized CO$_2$ under these conditions. Similar results for the SC-CO$_2$ deacidification of other vegetable oils have been previously reported for olive [102] and for roasted peanut [103] oils. At higher pressures (32.0 MPa), FFA concentrations were lower (around 20 wt.-%), what was probably due to the higher solvent selectivity for TG at higher pressures, which compositions achieved up to 72 wt.-% in the raffinate samples. Finally, the phytosterol contents, especially oryzanol, were 3 times higher than the ones found in commercial available high-oryzanol RBO.

Dunford and King [99] have also investigated the effects of isothermal and temperature gradient operation of a supercritical countercurrent column on the compositions of different RBO fractions. The 1.70 m height and 1.43 cm internal diameter packed column presented four different heated zones. Experiments were carried out during 180 minutes at 20.5 MPa, with temperatures varying from 40 to 90 °C and, once more, no extract reflux was employed. Feed composition was the same as presented above (Table 3.2).

Application of temperature gradients along the column proved to be beneficial in reducing TG losses in the extract samples, improving the purity of the extract through an "induced" internal reflux within the column. Utilization of higher temperatures in the stripping section improved FFA removal from crude RBO. The best results for the removal of FFA in the extract fraction (52 wt.-%) were obtained by employing 60-70-80-90 °C gradient temperature profile. Higher temperature zones at the top of the column reduced the CO_2 density, causing the condensation of the less volatile components, inducing then the internal reflux. Low-volatile components were then enriched in the raffinate fraction: TG (from 70.7 to 73.2 wt.-%), oryzanols (up to 1.41 wt.-%) and sterol esters (up to 4 wt.-%). The FFA content in the raffinate samples was reduced up to 4.4 wt.-% when applying a 45-55-65-75 °C gradient temperature profile. By using the mentioned approach, the authors have also obtained RBO fractions with a total sterol ester content (23 % HPLC area) higher than that of commercially available sterol-ester enriched margarines/spreads (21 % HPLC area), what proved the efficiency of using SC-CO_2 in the enrichment of low volatile components in the raffinate fractions.

Supercritical fluid fractionation was also investigated as an alternative method to recover phytosterol-enriched TG fractions from vegetable RBO deodorizer distillates (DD) [101]. The fractionation experiments were performed at a constant pressure over the pressure range of 13.6–27.2 MPa with the possibility to use gradient temperature profiles in the same extraction column previously described [99]. FFA were the main components (30–40 wt.-%) in the DD and the majority of the phytosterols were present in the free form (10–20 wt.-%). Although a significant portion of the oryzanol present in crude RBO was lost during the conventional oil refining process, the authors were not able to detect oryzanol in the RBO DD. A two step SC-CO_2 fractionation method was developed to obtain low FFA and high sterol content oil fractions from RBO DD. Initially, the FFA content of the DD was reduced to less than 10 wt.-

% at a relatively lower pressure and temperature, 13.6 MPa and 45 °C. The first extraction step was designed to remove the FFA, since these compounds are not desirable in edible oils. A sterol ester-enriched TG fraction (20 % sterols, 38 % TG) could be collected at a higher pressure, 20.4 MPa, by utilizing a second extraction step. FFA concentrations were about 5 wt.-%. The second extraction step further improved the product quality because some of the color-producing compounds in the DD, which remained in the system during the first SC-CO$_2$ extraction, could be then separated from the final product.

A two-stage columnar SC-CO$_2$ fractionation process for the enrichment of phytosterols and their respective esters in vegetable oils, particularly vegetable oils containing oryzanol and/or fatty acid or ferulic esters of phytosterols (like RBO, corn fiber and soybean oils, and vegetable oils DD) was proposed by Dunford et al. [70]. One year later, this process was patented in the USA [100]. A simplified flowsheet of the proposed process can be visualized in Figure 3.9.

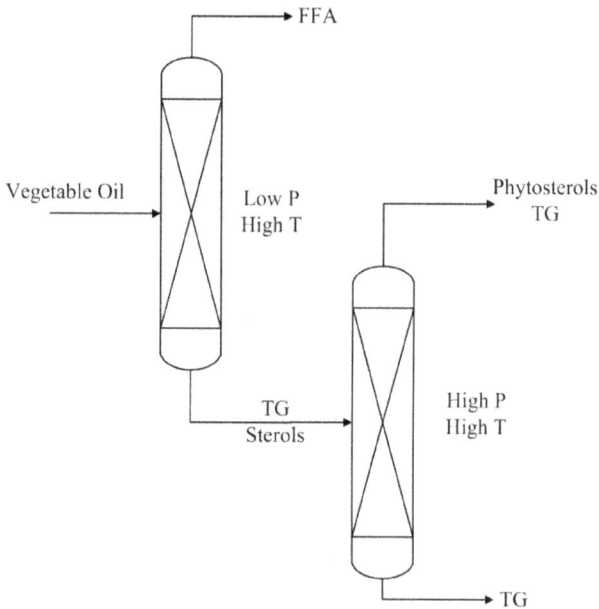

Figure 3.9. Schematic diagram of a two-step fractionation process for phytosterol ester enrichment in vegetable oils [70].

Initially, FFA was removed in the extract fraction at high temperature and low pressure (13.8 MPa and 80 °C). Then a phytosterol-enriched oil fraction was obtained with a second step extraction process, where the raffinate obtained in the first process step was introduced into the column as feed material, achieving then a good separation between TG and phytosterols. The extract from the second fractionation step can then be used as a functional and/or nutraceutical food ingredient. Using this method, phytosterol contents of 5 wt.-%, 7.5 wt.-%, 10 wt.-% or even 15 wt.-% can be obtained in the extract stream. The resultant raffinate is a low acidity TG-rich edible oil.

4. Mandarin Orange (*Citrus reticulata*)

The awareness of the health benefits associated with the consumption of fresh fruit has led to an increase in demand for citrus fruits, in particular mandarins and tangerines, which ones possess their own distinct flavors and aromas.

Mandarin belongs to the genus *Citrus* which includes also lemons, tangerines, limes, grapefruits and several other important species. Mandarins were originally cultivated in China many centuries ago, spreading from China into Japan, India, Arabia and North Africa before reaching Europe. Only in the 19th century they were introduced to the colonies in America and Oceania [104].

Mandarins are cultivated worldwide between the latitudes 35° north and 23° south [105], with a production currently estimated to be in the range of 10-13 million tons per year [104]. Mandarin trees can grow under a variety of climates ranging from deserts to subtropical Mediterranean ones. Compared to other citrus-plants, the mandarin tree can reach a height of up to 7.5 m and withstand rather cold conditions. On the other hand, the fruit itself is quite cold sensitive.

The mandarin, *Citrus reticulata*, is the largest and most varied group of edible citrus, what makes its taxonomical classification a very difficult task. There is much confusion surrounding the differences between all available *Citrus* species and often the names are used interchangeably [104]. And this is considered an international issue: the "Ponkan" mandarin, for example, is the most common mandarin in Brazil, although it is also known as "Cravo" tangerine there. In China, the word "kan" is used to describe the larger, sweeter mandarins, while "chü" is used for smaller mandarins – both sweet and sour varieties. Indeed, same botanical varieties can be known by different names depending on the region it is cultivated. A good example is the "Ortanique", a Jamaican mandarin, successfully cultivated worldwide. In Australia it is known as "Australique", and in Cyprus as "Mandora" [104]. Tanaka [106] classified all different mandarin cultivars into more than 30 different species, while other researchers see them as a specific type of mandarin or even consider all members of the genus *Citrus* to be one single specie [107]. Sometimes the classifications like tangerine and clementine are also used to group different mandarin breeds. These taxonomical difficulties arise due to interfertility, similarity in phenotype, different hybrid forms, and lack of most wild forms after several years of cultivation. Traditionally, the classification relied on morphology [107], but nowadays mandarin varieties have also been classified by chemotype [108] and DNA markers [107].

The differentiation between citrus species and cultivars can be better evaluated when considering the chemical characterization of the fruits, flowers, as well as peel and leaf essential oils, what is about to be discussed in the next sections.

4.1. Mandarin Peel Oil (MPO)

Over 150 components have already been identified in mandarin and bergamot peel oils [109]. Main components of MPO are limonene and other terpenes, like γ-terpinene, corresponding to approx. 97 wt.-% of the oil. Oxygenated compounds like linalool, decanal, sinensal and methyl-N-methyl-anthranilate (MNMA) have the highest contribution to the aroma fraction and correspond to approx. 2 wt.-% of the oil. Important classes of high-boiling compounds found in mandarin oil are complex molecules, like flavonoids and waxes, at concentrations below 1 wt.-% [110, 111].

Terpenes are naturally occurring hydrocarbons. They mostly have very strong smells and are derived from units of isoprene, $CH_2=C(CH_3)CH=CH_2$, joined together into chains and rings. The name is derived from turpentine, a liquid consisting of several terpene components distilled from the resin of pine trees. Terpenes can be classified in several groups, such as monoterpenes ($C_{10}H_{16}$), which ones constitute the major emissions from conifers and citrus trees, and sesquiterpenes ($C_{15}H_{24}$), also commonly found in citrus trees. The usual terpene nomenclature and some examples are presented in Table 4.1.

Table 4.1. Usual nomenclature of terpenes and examples.

Number of C-atoms	Nomenclature	Examples
10	Monoterpene	Limonene, pinene, sabinene, terpinene
15	Sesquiterpene	Caryophyllene, farnesene, humulene
20	Diterpene	Cembrene, taxadiene
25	Sesterterpene	Haslenes
30	Triterpene	Squalene
40	Tetraterpene	Carotenes (alpha-, beta-, gamma-), lycopene
> 40	Polyterpene	Natural rubber (latex)

Terpenes are the primary constituents of the essential oils of many types of plants, flowers and fruits and are used widely as natural flavor additives for food, as fragrances in perfumery, in aroma therapy, in detergents and soaps formulations and in traditional and alternative/folk medicines. Table 4.2 shows the structures of some common monoterpenes. Due to the high concentrations of nonpolar terpenes, especially limonene, MPO is practically insoluble in water. Limonene and other monoterpenes are very sensitive, especially when heated or under UV-radiation in the presence of oxygen. Therefore, high concentrations of monoterpenes are completely undesired: they are responsible for limiting the oil solubility in normal beverages, its use in bakery products, reducing also the shelf-life of food products [111]. Terpenes can be removed from the crude oils without changing their flavors significantly, a process which is called deterpenation.

Table 4.2. Chemical structures of monoterpene components.

p-Cymene p-isopropyl-toluene Molecular formula: $C_{10}H_{14}$ Molecular weight: 134.2	**Camphene** 2,2-Dimethyl-3-methylene- bicyclo[3.1.1]hept-2-ene Molecular formula: $C_{10}H_{16}$ Molecular weight: 136.2
Limonene (-)-Carvene (S)-4-Isopropenyl-1-methyl cyclohexene (-)-p-Mentha-1,8-diene Molecular formula: $C_{10}H_{16}$ Molecular weight: 136.2	**Myrcene** 7-Methyl-3-methylene-1,6- octadiene Molecular formula: $C_{10}H_{16}$ Molecular weight: 136.2
α-Pinene (1S,5S)-2-Pinene Molecular formula: $C_{10}H_{16}$ Molecular weight: 136.2	**β-Pinene** (1S,5S)-2(10)-Pinene Molecular formula: $C_{10}H_{16}$ Molecular weight: 136.2
Sabinene (1R,5R)-4(10)-Thujene Molecular formula: $C_{10}H_{16}$ Molecular weight: 136.2	**γ-Terpinene** p-Mentha-1,4-diene 1-Isopropyl-4-methyl-1,4- cyclohexadiene Molecular formula: $C_{10}H_{16}$ Molecular weight: 136.2

The oxygenated components presented in citrus oils are responsible for their typical flavor characters. Some of more than 200 aroma components are present at very low concentrations, in the range of ppm or even below detection limits. Chemical structures and some characteristics of oxygenated components present in MPO are summarized in Table 4.3.

Table 4.3. Chemical structures of oxygenated components present in MPO.

Carvone *p*-Mentha-6,8-dien-2-one (S)-5-Isopropenyl-2-methyl- 2-cyclohexenone Molecular formula: $C_{10}H_{14}O$ Molecular weight: 150.2		**Thymol** 2-Isopropyl-5- methylphenol 5-Methyl-2- isopropylphenol Molecular formula: $C_{10}H_{14}O$ Molecular weight: 150.2	
Camphor 1,7,7-Trymethyl-bicyclo heptan-2-one Molecular formula: $C_{10}H_{16}O$ Molecular weight: 152.2		**1,8-Cineole** Eucalyptol 1,8-Epoxy-p-menthane Molecular formula: $C_{10}H_{16}O$ Molecular weight: 152.2	
Linalool (±)-3,7-Dimethyl-1,6- octadien-3-ol (±)-3,7-Dimethyl-3- hydroxy-1,6-octadiene Molecular formula: $C_{10}H_{18}O$ Molecular weight: 154.3		**α-Terpineol** (S)-*p*-Menth-1-en-8-ol (S)-2-(4-Methyl-3- cyclohexenyl)-2- propanol Molecular formula: $C_{10}H_{18}O$ Molecular weight: 154.3	
Decanal Caprinaldehyde Decyl aldehyde Molecular formula: $C_{10}H_{20}O$ Molecular weight: 156.2		**Eugenol** 4-Allylguaiacol 4-Allyl-2- methoxyphenol Molecular formula: $C_{10}H_{12}O_2$ Molecular weight: 164.2	
Methyl-N-Methyl- Anthranilate (MNMA) Anthranilic acid Molecular formula: $C_9H_{11}NO_2$ Molecular weight: 165.2		**Sinensal** 2,6,10-trimethyl- 2(E),6(E),9(E),11- dodecatetraenal Molecular formula: $C_{15}H_{22}O$ Molecular weight: 218.3	

From the above mentioned oxygenated aroma components, thymol, linalool, terpineol, decanal, MNMA and sinensal present relative high concentrations in MPO. Aldehydes (like decanal, octanal and sinensal) are very important components to the flavor of citrus oils, being also used as an indicative of the product´s quality [111, 112].

The characteristic smell of mandarins and tangerines is often mentioned to be caused by thymol and MNMA, but this proposal has never been fully proven. According to gas chromatography-olfactometry analysis, additional components, such as γ-terpinene and β-pinene, have been proposed to be necessary in the generation of the typical peel oil flavor [113]. Also, the relatively high quantities of linalool and sinensal in mandarins have been considered as the main factor for the typical MPO flavor. Additionally, the tangerine-like smell was suggested to be mainly based on carbonyl compounds, such as α-sinensal, decanal and perillaldehyde [113].

Table 4.4 shows a comparison between the composition profiles of different citrus oils. It is possible to observe that oranges and mandarins present the highest terpene concentrations (especially monoterpenes), while the other oils present higher concentrations of oxygenated components. MPO presents the richest concentration profiles, once their compositions can vary drastically from ppm to wt.-%, depending especially on the cultivar chosen.

Table 4.4. Composition of different citrus oils (adapted from [112]).

Citrus specie	Terpenes	Oxygenated compounds
Orange	96 % limonene	1.6 % aldehydes
Grapefruit	89 % limonene	1.8 % aldehydes
Lemon	65 % limonene 10 % pinene 10 % terpinene	up to 13 % citral
Bergamot	32 % limonene 6 % pinene 7 % terpinene	12 % linalool 30 % linalyl acetate
Mandarin	52-96 % limonene up to 2 % pinene up to 36 % terpinene	up to 1.2 % aldehydes up to 1.2 % linalool 1.0 % MNMA 0.2 % Sinensal

The composition of mandarin oils and its dependency to cultivar and origin has been analysed by many authors using GC [108, 109, 113-119] and LC [120]. Good reviews on mandarin oil compositions were published by Lawrence [120-122], highlighting the progresses in the field in the last years.

Lota et al. [109] investigated 41 different cultivars of *Citrus reticulata* Blanco and identified 63 different compounds. Cultivars were grouped in two different chemotypes: the first one, rich in limonene (83.8-96.2 %) and containing up to 6.0 % γ-terpinene; the second one contained only 52.2-81.3 % limonene and 11.2 to 36.7 % γ-terpinene. Table 4.5 shows the concentration profiles (in wt.-%) of 7 different mandarin cultivars, identified as follows: Palazzelli (no.1), Cravo (no.2), Osceola (no.3), Ponkan (no.4), África do Sul (no.5), Imperial (no.6) and Federici (no.7).

Table 4.5. Chemical compositions of MPO from different cultivars (adapted from [109]).

Components	Mandarin cultivars						
	1	2	3	4	5	6	7
α-Pinene	0.3	0.5	0.4	0.9	0.3	2.1	1.0
Sabinene	0.2	0.7	1.3	0.3	0.2	1.0	0.3
Myrcene	1.6	1.7	1.6	1.7	1.4	1.7	1.3
Limonene	96.2	95.1	92.5	89.2	88.8	70.2	52.2
γ-Terpinene	tr	tr	0.5	4.6	4.5	17.4	36.7
Terpinolene	-	-	tr	0.2	0.2	0.8	1.7
Decanal	0.1	-	0.1	0.1	0.4	0.1	0.1
Linalool	0.2	0.6	0.8	0.7	1.2	0.2	0.3
α-Terpineol	tr	tr	-	0.1	0.1	0.2	0.4
MNMA	-	-	-	-	-	-	1.1
Thymol	-	-	-	-	-	0.5	-
Sinensal[a]	0.1	0.2	0.1	tr	tr	-	0.1
Others	1.0	1.2	2.7	2.2	2.9	5.8	4.8

[a] – corresponds to the sum of α- and β- isomers; tr – traces.
Varieties: Palazzelli (1), Cravo (2), Osceola (3), Ponkan (4), África do Sul (5), Imperial (6) and Federici (7).

The 63 identified components accounted up to 99.6 wt.-% of the total amount of the oil. Peel oils consisted almost exclusively of hydrocarbons with limonene as major component (52.2 to 96.2 %), associated with γ-terpinene (traces up to 36.7 %). α-Pinene (up to 2.1 %), linalool (0.1 to 2.5 %), myrcene (1.3 to 1.7 %) and sabinene (up to 1.3 %) were present in almost all samples investigated.

4.2. Conventional Peel Oil Extraction and Deterpenation Processes

After the juice extraction, the remaining residues are composed basically by peels and other solids, accounting to approximately 50 wt.-% of the fruits. The oil that remains in the peel must be extracted in order to recover very important components, which are normally treated as waste material. Citrus peel oil can also be recovered from the aqueous phase after juice expression [112, 114].

MPO extraction is usually performed by steam distillation or cold-pressing and the yields may vary from 0.21 (steam distillation) to 0.71 wt.-% (cold pressing) of the mandarin fruit itself [114].

Steam distillation is one of the most used extraction methods and is based on the different solubilities of the high valuable MPO components in the steam stream. This method is employed in industries mostly due to its simplicity and to the economical aspects involved, since significant amounts of raw material can be processed simultaneously. On the other hand, the main disadvantage of steam distillation extraction resides in the high operational temperatures used, what can be responsible for the degradation of thermolabile valuable components. In despite of that, cold-pressing has become the most used method for citrus peel oils extraction, since it can be carried out at low temperatures, achieving thus higher extraction yields, as previously mentioned [111, 112].

There are three different types of cold-pressed MPO available. They are distinguished by the color and maturity of the used fruits: the so-called "green oil" is obtained from immature fruit and mainly used as fragrance. The "yellow oil" is obtained from mature fruits, being used as flavor and fragrance, while the completely matured fruits are used to produce the "red oil", mainly used as flavor.

The deterpenation process consists mainly on the removal of limonene from the oil. As other monoterpenes are more volatile than limonene, they can be separated earlier or at the same time, improving the final quality of the oil, i. e., producing folded aroma concentrated mixtures. Depending on the amount of terpenes present in the oil, it is possible to concentrate the aroma fraction up to 20-fold or even achieve terpeneless concentrates [112].

Due to the presence of several thermolabile components, fractionation of the oil should be carried out at low temperatures. Since aroma components present very similar chemical and physical properties, multistage processes should be employed in order to provide a good separation. Conventional deterpenation methods include chromatographic procedures, adsorption/desorption, solvent extraction and vacuum distillation.

Adsorption/desorption on silica-gel can be chosen when terpeneless solutions are required. However, the total elimination of hydrocarbons is not required, since complete removal will affect the final properties of the oil [112, 123].

Conventional solvent extractions employ generally aqueous solutions of methanol, ethanol and iso-propanol. The similarities in the densities, as well as the removal of residual solvent constitute the major disadvantages of the method. Most extraction procedures were only tested in laboratory scale, with no further industrial scale-up investigations [111].

Vacuum distillation is the most used conventional deterpenation process [111, 112]. Based on the equilibrium data of the desired components, removal of terpenes can be achieved under vacuum conditions (below 1 kPa) at mild temperatures (usually between 40 and 60 °C). The process becomes difficult when the oil presents high concentrations of volatile aldehydes, like octanal and decanal, which ones can be removed along with the terpenes, especially due to their small boiling points and also to the modifier effect of limonene. This can be well exemplified by lemon peel oil. When applying the so-called ultra-high vacuum distillation (pressures as low as 0.002 kPa) for orange peel oil, an enrichment from 0.45 wt.-% to 90 wt.-% linalool could be achieved. Acetaldehyde from orange water-phase was also enriched from 0.08 wt.-% to 80 wt.-%.

Alternative methods for extraction and deterpenation of citrus oils are available in order to minimize or even reduce limitations of the traditional techniques, especially when scale-up procedures for industrial applications are desired. As examples, the following methods are mentioned: use of microwaves, laser photolysis, pervaporation and SFE. An overview on supercritical fluid applications to citrus oil deterpenation is presented as follows.

4.3. Applications of SCF to Citrus Peel Oils Deterpenation

4.3.1. Adsorption and Desorption Processes

Adsorption processes in the field of citrus oil deterpenation by means of SCF have been extensively investigated for approx. two decades. Cully et al. [124] published the first patent of a citrus peel oil deterpenation process with SC-CO_2. Different adsorbents were employed (silica gel, aluminium oxide, cellulose, bentonite, among others) at temperatures varying from 50 to 70 °C and pressures from 7.0 to 9.0 MPa, achieving up to 95 wt.-% terpene removal, highlighting then the main advantages of the process, such as high selectivity, production of special fractions of oxygenated (polar) components and the possibility to obtain products free of phototoxic and other high volatile components. Since the process can selectively fractionate products of high purity with very compact equipments, supercritical adsorption followed by desorption may be the method to be chosen. Indeed, the method is also extremely recommended when low quantities of raw materials have to be fractionated into products of high market value [112].

In the last years, several authors reported about their works on selective ad- and desorption with different citrus peel oils, as reviewed by Reverchon [10]. Table 4.6 summarizes some of these works, presenting also the employed process conditions.

Yamauchi and Sato [125] reported the fractionation of lemon peel oil using supercritical fluid chromatography on silica gel at 40 °C and with two pressure steps of 10 and 20 MPa, respectively. Ethanol was added as a modifier to remove the oxygenated aromatic components. They have observed that the amount of water present in the oil reduced considerably the activity of the adsorbent, concluding that the removal of water must be previously performed.

Table 4.6. Papers on selective desorption and pressure swing adsorption of citrus peel oils and corresponding model mixtures.

	Selective desorption				
	terpene desorption		'aroma' desorption		
Peel oil	P (MPa)	T (°C)	P (MPa)	T (°C)	Reference
Citrus, general	7-9	50-70	20-30	30-70	[124]
Lemon	7.5	40	8.5-12	40	[126]
Lemon	7.4	40	8.7	40	[127]
Mandarin Distilled Lime	7.5	40	8.5-12	40	[128]
Model mixture	7.5	40	20	40	[123]
	Pressure swing adsorption				
	adsorption		'aroma' desorption		
Peel oil	P (MPa)	T (°C)	P (MPa)	T (°C)	Reference
Citrus, general	8.8	40	19.4	40	[129]
Orange	8.8	40	19.4	40	[130]
Bergamot	8.8	40	24.8	40	[131]

The work of Barth et al. [127] investigated the separation between the terpene fraction and low-volatile components, such as coumarins, psoralens (phototoxic components) and waxes from lemon oil by different pressure levels with SC-CO_2. Silica gel was used as adsorbent and the experiments were conducted at 40 °C. Monoterpene hydrocarbons were obtained at the lowest pressure level (7.5 MPa), while deterpenated aroma concentrates without low-volatile components and waxes were fractionated at 8.7 and 11.5 MPa, respectively. Approximately 10 wt.-% oxygenated components was lost with waxes, while 3 wt.-% was lost during the terpene fractionation.

Reverchon [123] employed SC-CO_2 in order to desorb limonene, linalool and their mixtures (50 and 80 wt.-% limonene) from silica gel to simulate the deterpenation of citrus peel oils. The influence of pressure, temperature, solvent flow rate and solute loading was studied. The maximum selectivity was obtained by operating at 40 °C and 0.1 kg/kg loading. The optimal desorption conditions were obtained when using two successive pressure steps: the first step performed at 7.5 MPa promoted the selective desorption of limonene and the second one, assured the fast desorption of linalool at 20 MPa. Adittionally, a Langmuir-like equation provided the best representation of the experimental data for each single component and for the mixtures. The desorption process was successfully modeled for the single components and also for limonene-linalool mixtures, with fairly well fitted data.

Alternatively to charging the peel oil onto the adsorbent at ambient conditions, it can be first dissolved in SC-CO_2. This is an advantageous way to remove undesired heavy-weight components as they are much less soluble and will stay in the mixing vessel. Then the loaded CO_2 is charged on the adsorbent and the oxygenated fraction is preferably adsorbed. In a second step this product fraction is desorbed with pure CO_2 of higher pressure. This process, called pressure-swing-adsorption (PSA), can be carried out continuously if at least two adsorbers exist. The oil is fed continuously because one of the adsorbers is always in adsorption mode while the other one is desorbed at reversed flow direction. Although the feed flow will be continuous and thus much easier to automate the aroma and terpene fractions will not accrue uniformly over the halfcycle. In order to increase the selectivity of the process, the bed can be rinsed with pure CO_2 at adsorption conditions before increasing the pressure [129, 130]. Thus, most of the terpenes that were adsorbed along with the aroma fraction will be desorbed similar to the first step in selective desorption. Besides PSA, it is also possible to perform a chromatographic separation after dissolving the oil in SC-CO_2, as presented by Pitol-Filho [132].

Reverchon et al. [133] and Subra et al. [21] have investigated the adsorption of a mixture composed by 13 terpenes (α- and β-pinenes, myrcene, limonene, γ-terpinene, β-caryophyllene, citronellyl- and geranyl acetates, linalyl acetate+geraniol, linalool, citronellal and citral) on silica gel with CO_2. These components have been divided into four pseudo-components according to the similarity of their behavior during the adsorption process. An equilibrium model was developed and the breakthrough time and the maximum concentration of the four pseudo-components were satisfactorily well modeled. Aditionally, they observed that a low temperature will increase selectivity, as the difference in adsorption capacities of the different components increases with decreasing temperature.

During the adsorption and desorption processes, residues of heavy components can remain adsorbed on the silica gel. The regeneration of the adsorbent can be performed by washing it with warm ethanol [127]. When PSA is applied most waxes will not reach the column and a steady-state will be reached. This was observed after 13 halfcycles for bergamot oil deterpenation [131].

The deterpenation and bergaptene removal from bergamot essential oil have also been performed by Araújo and Farias [134]. The characterization of the raw material used in the experiments was not presented in the paper. Limonene was firstly desorbed from silica gel at 40 °C and 7.7 MPa. In the second step, bergaptene was separated (92.7 wt.-%) from the other oxygenated components at 50 °C and 15.1 MPa. The final concentrations of linalool and ethyl acetate achieved approx. 52.0 and 69.0 wt.-%, respectively.

4.3.2. Countercurrent Fractionation Processes

In order to avoid partial removal of volatile oxygenated components with the terpene fraction, a multistage countercurrent process is necessary. It is well known that the separation of terpenes (especially monoterpenes) from low volatile oxygenated components can be easily achieved, but multistage countercurrent processes produce very interesting results when very high concentrated raw materials are used. Citrus oils are one example of these complex mixtures, with total terpene concentrations up to 99.0 wt.-%. Therefore, investigations of the operational parameters on the recovery of determined components and improvement of product purity must be performed [112].

Once the geometrical characteristics of the process are defined, namely: determination of column height, inner diameter and type of packing material to be used, experimental operational conditions, like temperature, pressure, solvent-to-feed and reflux ratios, liquid and vapor cross section capacities must be defined. These conditions can be chosen through analysis of the equilibrium data available for each specific system [112]. The determination of the number of equilibrium stages for one separation process can then be performed as described in Chapter 2.3. Table 4.7 summarizes some works available in the literature for the countercurrent multistage fractionation of citrus oils or even model mixtures composed by their most representative components with SC-CO$_2$.

Table 4.7. Published studies on countercurrent SFE of citrus oils with SC-CO$_2$.

Systems investigated	Operational conditions	References
Different citrus oils	7.0-9.0 MPa; 55-75 °C	[135]
Bitter/sweet orange peel oils	8.0 MPa; temperature gradients 60, 85 and 75 °C	[136]
Eucalyptus essential oil (system limonene-1,8 cineole)	Up to 9.0 MPa; 40 and 45 °C	[137]
Model mixture (60% limonene, 10% γ-terpinene, 20% linalool, 10% linalyl acetate)	7.5 to 8.0 MPa; 40-80 °C	[138]
Orange oil + model mixture 80 % limonene and 20% linalool	8.8 and 9.8 MPa; 50 °C	[129]
Orange peel oil	8.8 MPa; 60 °C and temperature gradients (40-60 °C)	[139]
Orange peel oil	8.0 to 13.0 MPa; 50 to 70 °C	[112]
Bergamot oil (model mixture limonene linalyl acetate)	8.8 MPa; 60 °C Computational simulations	[140]
Bergamot essential oil	Up to 9.0 MPa; different temperature gradients	[141]
Lemon peel oil	8.4 to 10.5 MPa 42, 50, 60 °C	[142]

Gerard [135] was one of the first authors to investigate the applicability of countercurrent deterpenation of citrus oils employing SC-CO$_2$. Internal reflux was proposed, once the column was heated with different temperature gradients, in order to increase the terpene concentrations in the extract stream. Internal reflux is usually applicable for laboratory scale columns of small diameter (maximum solubility differences caused by isobaric vapor-phase behavior), but the process can be better controlled when using external reflux, which is limited only by economical aspects [112]. As expected, the bottom products (raffinate) was composed mainly by carotenoids and low volatile components (oxygenated aroma components). One suggestion proposed to improve the separation of components with very similar vapor pressures was the change of the polarity of the solvent by adding water as an entrainer. In this case, the chosen components were anethole and caryophyllene, which present different solubilities in CO_2 only when saturated with water, enabling then the separation of sesquiterpenes from oxygenated components.

Stahl et al. [136] treated sweet and bitter orange peel oils at 8.0 MPa. The starting feed material presented 90 wt.-% hydrocarbon terpenes. By applying different axial temperature gradients, i. e., internal reflux, they intended to achieve smaller losses during the deterpenation process. The column displayed the highest temperature in its middle, 85 °C, while the temperatures were fixed as 75 and 60 °C at the top and bottom of the apparatus, respectively. Lower solvent density (166.7 kg/m^3) was obtained at the top, what provided a better separation between terpenes and oxygenated components: less than 1 wt.-% oxygenated components were obtained as top product, while the concentration of hydrocarbons at the bottom reached 42 wt.-%.

The mass transfer efficiency in the separation of a mixture of two terpenes (limonene and 1,8 cineole) through countercurrent fractionation with CO_2 was performed by Simões et al. [137]. The overall mass transfer coefficient was predicted by employing different models. The values obtained were relatively high with comparison to experimental data, obtaining then separation factors close to 1, what is considered insufficient for a good process separation. It is also important to mention that the column employed was only 1 m height.

A model mixture composed by 60 wt.-% limonene, 10 wt.-% γ-terpinene, 20 wt.-% linalool, 10 wt.-% linalyl acetate was employed as feed material in the countercurrent fractionation with SC-CO_2 [138]. The investigated operational conditions ranged from 7.5 to 8.0 MPa and 40 to 80 °C. Limonene separation was slightly improved when the reflux ratio was very high, tending to infinite. Without reflux, γ-terpinene was identified in the extract stream, mainly because it presents a lower volatility than limonene. On the other hand, its separation from limonene was successfully achieved when applying total reflux. At 8 MPa and total extract reflux, extract concentrations were maximized when the temperatures were increased from 70 to 80 °C, reaching higher solubility conditions. As a conclusion, the countercurrent deterpenation can provide very interesting results when performed at higher temperatures, since the thermal stability of the products is not affected.

Cold-pressed orange oil from Brazil and a model mixture of 80 wt.-% limonene and 20 wt.-% linalyl acetate were investigated through total reflux and continuous countercurrent operations by Sato et al. [139]. For a total reflux operation, CO_2 flow rate was most important to remove the terpene fractions. The increase of solvent flow rate or solvent-to-feed ratio produced a higher selectivity between terpenes and oxygenated components. When orange oil was used as feed material, the selectivity increase was smaller than that of the model mixture. For a continuous operation with temperature gradients (40-60 °C) at 8.8 MPa, the authors have found that terpenes were enriched at the top and oxygenated components and waxes were better separated at the bottom and in the side stream, respectively. The side stream was used in order to avoid a higher concentration of waxes in the raffinate fraction, what could cause further purification problems when separating them from the desired oxygenated components.

The work of Budich [112] is one of the best works published so far on the countercurrent deterpenation of citrus oils using SC-CO_2. The separation of terpene hydrocarbons from oxygenated aroma components was investigated through the fractionation of orange peel oil. In order to provide a complete understand of the system behavior, phase equilibria measurements, countercurrent column experiments and flooding point measurements were carried out. Experimental conditions evaluated ranged from 50 to 70 °C and from around 8.0 MPa up to 13.0 MPa. Separation factors for the two groups of oil components, *terpenes* and *aroma*, were obtained for the system CO_2-orange peel oil at different temperatures and pressures. Short-cut stage calculation method based on the Jänecke diagram was applied to evaluate equilibrium data. By performing these calculations, the height of a theoretical stage was found to be 0.5 m. At 60 °C, 10.7 MPa and a solvent-to-feed ratio of 100, 18 theoretical stages (9 m height column) and a reflux ratio of 2.5, were required to produce a 20-fold concentrate containing 69 wt.-% terpenes from a feed material concentration of 98 wt.-% terpenes. The extract fraction was composed of 99.8 wt.-% terpenes, achieving then the desired high purity. With a feed flow of 100 kg/h, the minimum inner diameter of the column determined from the flooding point data would be 0.4 m, proving that, since the objective is to produce flavor fractions of a high market value, supercritical countercurrent deterpenation can be a very competitive process.

Kondo et al. [140] developed a process flow diagram for citrus oil processing with SC-CO_2, employing an extraction column with/without reflux in order to evaluate the separation performance of a model mixture composed of limonene (20 to 80 wt.-%) and linalyl acetate. For the calculation using the simulator SIMSCI PRO/II at 60 °C and 8.8 MPa, the effects of operating conditions on extraction ratio of limonene, separation selectivity, and recovery of linalyl acetate could be observed as a function of the solvent-to-feed ratio. They have also investigated the best feed inlet positions. They concluded that extract reflux was not a rewarding technology for this case. The performance was improved with increase in the stage number at higher solvent-to-feed ratios.

The efficiency of separation of bergamot essential oil carried out in a countercurrent column filled with Raschig rings and using SC-CO_2 as solvent was performed by Poiana et al. [141]. In the experiments carried out, the direct effect of CO_2 density and the solvent-to-feed ratios were investigated. The conditions that produced extracts with a similar volatile fraction composition of starting material, composed mainly by limonene (32.1 wt.-%), linalool (12.1 wt.-%) and linalyl acetate (29.7 wt.-%), and with a high yield (more than 80 % of recovery) were those with a low solvent-to-feed ratio. The lowest bergaptene content was obtained at low solvent density or at high solvent-to-feed ratio. Once more, temperature gradients were employed in order to achieve internal reflux. The best result was obtained at a CO_2 density of 206 kg/m^3 (8 MPa and a temperature gradient of 46-50-54 °C) and a solvent-to-feed ratio of about 9.5; in this separation a bergaptene content lower than 0.01 wt.-% could be measured.

5. Analytical Methods and Experimental Set-Ups

5.1. Analytical Methods

Different analytical methods were employed in order to analyze the samples obtained. The most used method in this work was gas chromatography (GC), especially because all fat components from RBO and terpenes and terpenoids from MPO could be successfully detected and analyzed. Reverse-phase High Performance Liquid Chromatography (HPLC) and gas chromatography-mass spectrometry (GC-MS) were performed by the central laboratory at TUHH, in order to verify the oryzanol content of RBO samples and for qualitative analysis of MPO samples, respectively. A brief description of the analytical methods employed in this work is presented in the sections below.

5.1.1. RBO Sample Analysis

RBO components investigated in this work – FFA, sterols and TG – were analyzed by gas chromatography. Oryzanols were identified by reverse-phase HPLC.

Gas chromatographic analysis were performed by using a HP 5890A GC. For the determination of FFA and sterols, a 30 m length DB-5 column (J&W Scientific) with an inner diameter of 0.25 mm and 0.1 μm film thickness was used. The program used was the following: oven temperature 70 °C, a temperature increase of 25 °C per minute up to 250 °C, then an increase of 4 °C per minute up to 350°C and finally 15 min at 350 °C. The injected samples were 1.0 μl and detector and injector temperatures were 360 and 320 °C, respectively. The TG analysis were performed using the same GC equipment with a 6 m length column (SGE, Australia). Other characteristics of the chromatographic column were inner diameter of 0.53 mm and 0.1 μm film thickness. The temperature program was as follows: oven temperature 60 °C (2 min), a temperature increase of 10 °C per minute up to 370°C, finishing after kept 5 min constant at 370 °C. The injected samples were 1.0 μl and detector and injector temperatures were 370 and 350 °C, respectively. Peaks of the most important components were identified by comparison to injected standards (analytical grade).

The quantitative analysis were performed through the relative method of the internal standards (ISTD). For the determination of FFA, sterols and TG concentrations, squalan was used as internal standard. RBO samples (20-30 mg) were diluted in 1 ml of a standard solution (1.0 mg squalan/ml solution), which was prepared with n-hexane and acetone (1:1 vol/vol), as described by Gottschau [143] and Machado [64], for the analysis of FFA and sterols. When analyzing TG, ethyl acetate was used as solvent for the preparation of the standard solution (1.5 mg squalan/ml ethyl acetate). Each sample (10 mg oil) was previously diluted in 1 ml solution, being then diluted up to 1:20 (vol/vol).

Additional to the internal standard method, the so-called Response Factors (RF) were determined. Sample compositions could only be quantitatively analyzed when peak areas were corrected by using RF for the most important components presented in the oil.

Peak area A_i is directly proportional to the concentration c_i of a substance, but not all substances correspond to the same peak area at a determined concentration. Therefore, RF of a component i is defined as:

$$RF_i = \frac{c_i \cdot A_{IStd}}{A_i \cdot c_{IStd}}$$ (5.1)

Response factors for the analyzed components account for the sensitivity of the detector for different molecular structures and should be repeated periodically, in order to make sure that deviations from the detector signal will not affect the final results.

The ISTD method is generally used as an analytical method because it takes into account errors and deviations caused by different injection volumes, being then considered a reliable method. Finally, the concentration of component i can be calculated as:

$$x_i = RF_i \frac{A_i \cdot c_{IStd}}{c_i \cdot A_{IStd}}$$ (5.2)

Response factors were determined for all RBO components evaluated, except for oryzanols. The obtained RF values are summarized and presented in Appendix A.

For the analysis of oryzanols, reverse-phase HPLC was employed. The method used was a modification of one method available in the literature [74]. Samples were prepared as follows: 0.7 g oil were diluted in 3 ml acetonitrile at 70 °C and kept under agitation for 20 minutes. After that, the mixture was centrifuged (16000 G) and 20 µl of the supernatant was collected and used for the analysis. A Superspher 125 (4 mm) column was employed and methanol was used as eluent at room temperature. The solvent flow rate was 0.7 ml/min (8.2 MPa) and the wavelength was fixed in 295 nm after previous tests with standard solution (20 mg oryzanol/l acetonitrile).

5.1.2. MPO Sample Analysis

Samples of raw material, extract and raffinate were analysed quantitatively by gas-chromatography. Approximately 10 mg of oil sample were diluted with 1 ml ethanol and injected into the GC (Hewlett Packard 5890). A DB-5 column (J&W Scientific) with an inner diameter of 0.25 mm and a film thickness of 0.1 µm was used. The carrier-gas was nitrogen (grade 5.0) and a split ratio of 1:50 was used. The GC was equipped with an HP 7673C autosampler and a flame ionization detector (FID) using hydrogen and synthetic air. Peak areas were calculated and recorded by an integrator (HP 3396B) and assumed equivalent to mass percent of the respective compound as described previously by Budich [112]. No internal standards were used, i. e., all response factors were set to 1.0. Due to very similar retention times of some MPO components, the temperature program was adjusted to the following one in order to achieve satisfactory peak separations: 5 min at 40 °C, then a temperature increase of 10 °C per minute up to 320 °C and finally 20 min at 320 °C. The injected samples were 1.0 µl and detector and injector temperatures were 350 and 320 °C, respectively. Peaks of the most important components were identified through comparisons of retention times of previous injected standards.

GC-MS analysis were carried out on a HP 5890II GC with a MSD 5971A detector in order to identify a larger number of components and perform a complete profile evaluation of the oils available for this work. Samples were diluted with 1.0 ml of dichloromethane and 0.25 µl of the resulting solution was injected into the GC-MS equipment. Basically, the temperature program was adjusted as follows: 3 min at 50 °C, then a temperature increase of 30 °C per minute up to 250 °C and keeping it finally constant for 1 min. The capillary column used was a RESTEX XTI-5 capillary column (95 % dimethyl and 5 % diphenyl polysiloxane), 30 m in length, 0.25-µm i.d., and 0.25-µm film thickness. Further information are available in the previous work of Smith et al. [144].

5.2. Experimental Apparatus and Procedures

5.2.1. Batch Extractions

When the amount of extract required for both raw materials (RBO and MPO) was low, i. e., for analysis, desorption and modeling purposes (up to 25 g of raw material), the extraction experiments were performed in a single-stage extraction apparatus, the Spe-ed[TM] SFE (Applied Separations, PA, USA). For MPO ad- and desorption experiments, a pilot plant (extractor volume up to 1.5 l) was assembled at TUHH and used in order to investigate the scale-up of the process. A simplified flowsheet of these batch extractors is presented in Figure 5.1.

The set-up of the batch extractors consisted basically of an oven/heated water bath, where the extraction column was placed. The column bottom was connected to the inlet valve where the solvent flowed through and the top of the column was connected to the outlet valve where the solvent flow rate was controlled and the extract collected. Outside the oven/water bath, the system pressure was reduced to atmospheric pressure during sampling and the total solvent flow rate could be measured by connecting rotameters to the system.

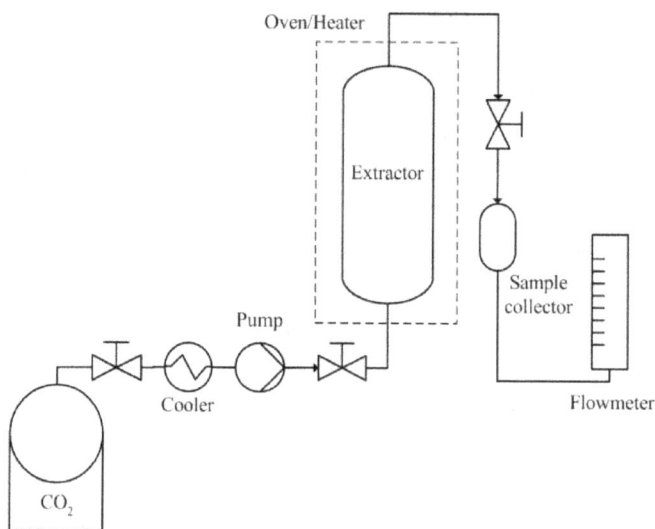

Figure 5.1. Simplified flowsheet of the batch extractors used.

Before starting the experiments, the pump cooler should be switched on in order to cool down the solvent and avoid cavitation problems. The fixed bed of particles was loaded and then the filled extractor placed into the oven. Extracts were then collected in weighted glass flasks by opening the outlet valve. By opening/closing the outlet valve, the average solvent flow rates were approx. 4.16 and 25 gCO_2/min for the Spe-ed[TM] SFE equipment and the pilot plant, respectively.

In order to achieve a large amount of extract (up to 2 kg/batch), what was necessary for the refining/deacidification RBO experiments, some batch extractions were performed with a larger batch extractor, which was composed basically by an autoclave with 15.8 l volume (maximum allowable pressure of 35 MPa at 150 °C), a separator (7.4 l volume) and thermostatic baths, operating with solvent recycle due to its high flow rates (up to 20 kg/h).

During the experiments it was observed that some rice bran particles – average diameter of 0.55 mm – were carried through the tubulations, causing clogging in some specific points. In order to avoid this problem, rice particles were agglomerated by using a laboratory drum agglomerator (Eirich, Germany). Water was added to the particles during the agglomeration procedure. The agglomerates obtained presented an average diameter in the range of 0.3-0.55 cm. After that, they were dried at 60 °C for approximately 24 h and stored until use.

5.2.2. Phase Equilibria Apparatus

The phase equilibria measurements were performed according to the static-analytical method employing an apparatus as presented in Figure 5.2 and previously described in the literature [145]. The apparatus was composed by a variable volume equilibrium cell maintained submerged in water. CO_2 was supplied by a pump (Maximator MSF 72-L01, Zorge, Germany) and the water temperature was controlled by a thermostatic bath. In order to maintain the pressure constant inside the cell when collecting samples, a water syringe pump (Varian aerograph 8500, California, USA) was coupled to the mobile piston located at the top of the equilibrium cell, which compensated the pressure losses, maintaining the pressure inside the equilibrium cell at the desirable levels, especially after collecting samples. The equilibrium cell presented also a stirrer to assure a good dissolution of the oil in the solvent and visualization windows, whose were useful in the identification of the phases and for loading/discharging oil procedures. The samples were collected from two different heights: a higher one, where the vapor phase was taken and a lower one, corresponding to the liquid phase. In order to control temperatures and pressures within the equilibrium cell, thermocouples and pressure gauges were employed.

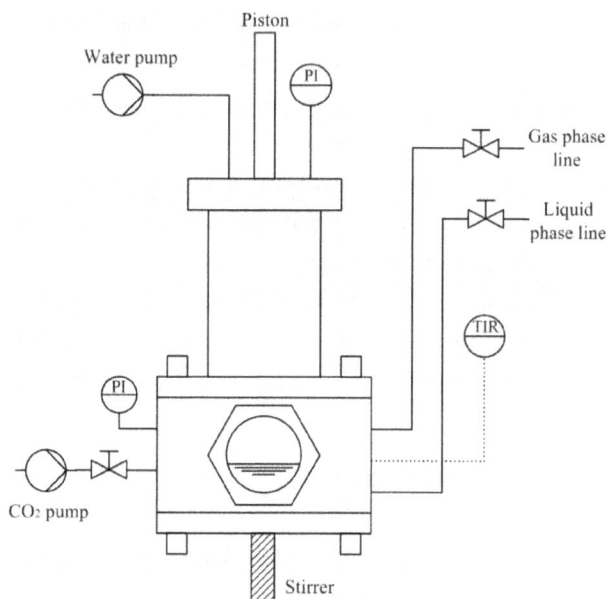

Figure 5.2. Scheme of the phase equilibria apparatus used.

Before collecting samples, equilibrium conditions must be achieved. Once the desired temperature and pressure were reached, the system was kept under agitation for approximately two hours. Thus, the stirrer could be turned off and after 30 minutes, the first sample could be taken. Samples were taken under vacuum conditions. A vacuum pump (Labovac P6D, Ilmenau, Germany) was coupled to the system, namely to the gas and liquid phase sampling lines. The amount of oil extracted was obtained gravimetrically and the amount of CO_2 was determined by applying a water volume system (composed by a burette system, see Figure 5.3) and a gas flow-meter. Gas volumes taken were chosen according to the operational conditions employed, varying from small (0.1 dm^3, liquid phase samples) to large (20 dm^3, gas phase) volumes.

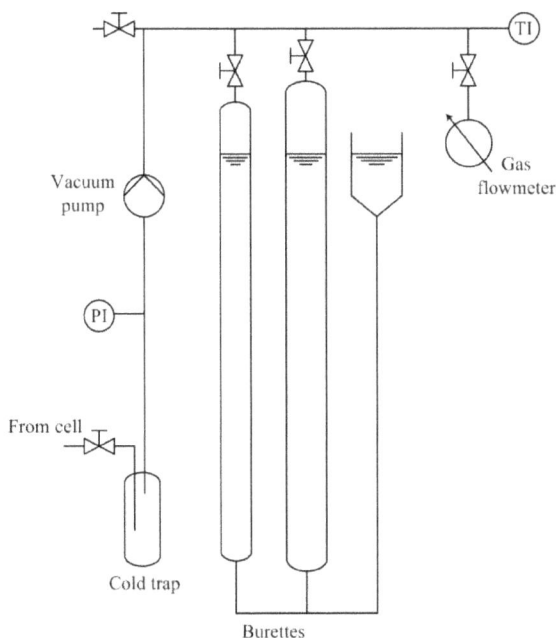

Figure 5.3. Burette system employed for the phase equilibria measurements.

5.2.3. Countercurrent Multistage Experiments

The countercurrent experiments were conducted in two different columns and a simplified flowscheme of both columns is presented in Figure 5.4.

The first column used presented a height of 6 m and 25 mm ID and was divided in four segments: two segments, each 2 m in length, filled with structured packing type EX (Sulzer Chemtech, Switzerland). The high-pressure section of the equipment was designed to operate up to 35 MPa and 150 °C and the low-pressure section to operate up to 10 MPa at the same temperature level.

A stainless stell tube (25 mm ID, ca. 0.7 m length) filled with stainless steel mesh packing (1.4571) was located above the enriching section in order to prevent the entrainment of droplets. Below the solvent inlet, a settling zone of 1 m was connected to the high-pressure section. Windows were placed in this section in order to control the raffinate level and for the observation of the falling droplets.

The experiments could be performed with extract reflux. Feed was normally supplied through the middle point of the high-pressure section by a piston pump (EKM, Lewa, Germany). All segments of the column and some connections were heated by means of electrical heating tapes (Horst, Germany) and heating plates. The temperatures were measured through Ni-CrNi thermocouples (type K, SAB Bröckskes, Germany) and the pressures by analogue and digital gauges.

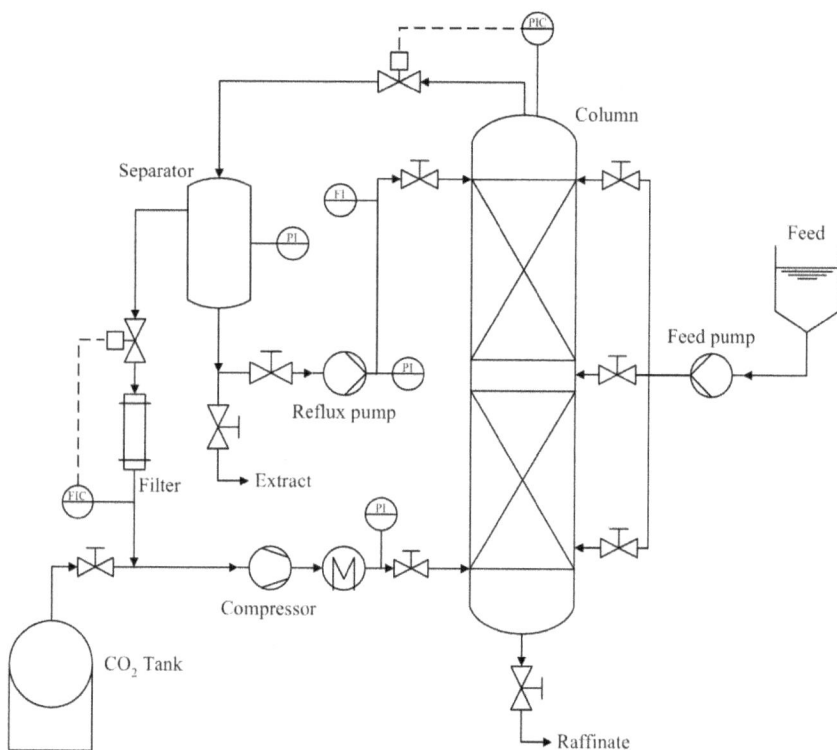

Figure 5.4. Flowsheet of the countercurrent apparatus used in the experiments.

Carbon dioxide was supplied at the bottom and was pre-heated to extraction condition before entering the column. A membrane pump (Burdosa, model MV-410, Germany) was employed in order to deliver the solvent. The pressure inside the separation column must be kept constant and it was controlled by an air-expansion valve (Kämer Ventile, Germany). After the valve, additional heating of the tubulations was required (Joule-Thomson effect).

The extract separation was performed in an empty vessel through pressure reduction and consequently expansion of the two-phase flow. The separator was heated by means of electrical heating tapes and the pressure controlled through another air-operated valve. When the pressure was considered too high (above 8 MPa), small quantities of solvent should be released manually through metering valves. Visualization windows placed at the bottom of the separator were used in order to keep the extract level constant when the experiments were carried out with reflux. By sampling, extract was withdrawn from its bottom and expanded to atmospheric pressure. Extract reflux was then charged to the top section of the column by means of a two- head piston pump (HK2, Lewa, Germany).

After the regeneration, the solvent was cooled down and pumped back to the system. Refrigeration was performed by a R22-driven cooler, which presented an independent loop of an alcoholic comercial solution that was pumped through this independent system by means of a centrifugal pump (Grundfos, Germany). The solvent and extract reflux flows were monitored/measured by Coriolis flowmeters (Rheonik, model RHM01, Germany), presenting in general an error lower than 1 wt.-%. Feed flow were determined by using a balance presenting a 0.1 g scale.

The second column consisted in an extraction column with a height of 7 m and 17.5 mm ID built from stainless steel high-pressure piping (316SS). The maximum operational conditions were 30.5 MPa and 110 °C. The height of the packings (Sulzer EX, Sulzer Chemtech, Switzerland) corresponded to 6 m and the column presented also packing free zones at the top and at the bottom (0.5 m each) for sample collecting purposes.

A visualization window was originally placed at the middle of the column in order to observe the flow regime inside the column. Flooding conditions could thus be detected visually. A second window was located at the bottom of the column (packing free zone), allowing the observation of the raffinate level.

The feed material was provided to the column at constant temperatures. A piston pump (type N-K 11, Bran & Lübbe, Germany) was responsible for the feed delivery and the flow rate set by changing the pump lift or the rate of revolutions (with a potentiometer). The feed supplying at three different column heights - 2.5 (bottom), 4.5 (middle) and 6.5 m (top) - was also possible. Depending on the case, three different operational modes could be employed: the co-current regime, (when the feed was supplied at the bottom of the column, very rarely used) and the counter-current regimes (feed supply at top and middle). Therefore, the column was divided into a stripping section below the feed height and an enriching one above that point.

Carbon dioxide was provided through a membrane compressor (MKZ 120-50, Andreas Hofer, Germany) and kept in a closed loop (recycling), being also electrically heated before entering the column. The extract flow at the top of the column was controlled by means of an air-driven regulating valve. By using this valve the column pressure could be kept constant during the extraction. When the extract reached the cyclone separator, the precipitation of the dissolved components occurred because of the pressure reduction.

The separator presented also a window at its bottom in order to allow the observation of the extract level. Thus, the extract could be collected by opening a valve at the bottom or it could be re-introduced into the column by a two-head piston pump (HK2, Lewa, Germany), working as a reflux pump. The carbon dioxide was then re-compressed to operational pressure.

Mass flow meters (type RHM01, Rheonik, Germany) and a second air-driven regulating valve located between the separator and the compressor controlled and adjusted the solvent flow rate. The entire equipment was electrically heated through heating tapes. The regulation was performed by two-steps controls and Pt-100 thermocouples. The temperatures within the column were determined with Ni-CrNi thermocouples located at five different column heights. The separator and reflux piping temperatures were also controlled by these thermocouples. In order to achieve good accuracy in the measurements, the temperatures, as well as the pressures, were determined by pressure couples within the column and were recorded by a regular PC equipped with an A/D converter card.

6. Deterpenation of Mandarin Peel Oil with Supercritical CO$_2$

In this Chapter, the results obtained for the deterpenation of MPO with SC-CO$_2$ through countercurrent multistage extraction and ad-/desorption are presented. Two crude MPO samples were investigated: a red oil from Spain and a green oil from Brazil.

First of all, the characterization of both raw materials will be presented, followed by considerations on phase equilibria measurements performed previously by several authors. Based on the vapor-liquid equilibria (VLE) data, the multistage extraction could be evaluated by employing the Ponchon-Savarit method, as described in Chapter 2.3.

Additionally, the oil samples available for this work were adsorbed in silica gel. The selective desorption of terpenes and aroma components from the adsorbent with SC-CO$_2$ was then investigated. By carrying out these experiments, optimum experimental conditions could be found, providing the best experimental conditions for the scale-up of the process.

6.1. Characterization of the Oil Samples

The red MPO used in this work was obtained from Spain (supplied by Sensient Essential Oils GmbH, Germany). Three different charges were used. The green MPO originated from Brazil was supplied by Duas Rodas Industrial Ltda. (Jaraguá do Sul, Brazil).

Qualitative composition was determined by GC/MS for the green and the third charge of the Spanish red oil. A total of 59 components were identified and are presented in Appendix B. In order to quantify the most representative components, GC analysis were performed as described in Chapter 5.1.2. The results obtained can be observed in Table 6.1. The Spanish oils presented a higher total terpene content, while the Brazilian green oil presented the richest aroma profile, explained mainly by the presence of three major components not found in the red oil samples: γ-terpinene, terpinolene and methyl-N-methyl-anthranilate (MNMA).

Table 6.1. Composition of MPO used in this work (wt.-%).

Compound	Red 1	Red 2	Red 3	Green
α-Pinene	0.499	0.495	0.446	2.049
Sabinene	0.988	0.586	0.553	1.798
Myrcene	1.688	1.747	1.677	1.680
Limonene	95.956	95.690	96.571	69.917
γ-Terpinene	-	-	-	20.036
Terpinolene	-	-	-	0.901
Total of terpenes	**99.131**	**98.518**	**99.247**	**96.381**
Linalool	0.406	0.222	0.246	0.139
Decanal	0.134	tr	0.132	0.299
MNMA	-	-	-	0.431
Sinensal	-	0.147	0.120	0.299
Total of aromas	**0.540**	**0.369**	**0.498**	**1.168**
Non-identified	0.329	1.113	0.255	2.451

The MPO components were grouped according to their retention times (RT) in the chromatographic column and according to comparisons with injected standards behavior and literature data. All the components identified up to approximately 11.6 minutes RT were defined as terpenes. Aroma components were identified and grouped between 12 and 22 minutes RT. Non- identified components were obtained beyond 22 minutes RT. They can be probably a mixture of waxes, flavonoids and other high boiling components, such as coumarins.

In order to compare visually both raw materials used, the chromatograms obtained are presented in Figures 6.1 and 6.2. Figure 6.1 shows the chromatogram obtained for one Spanish oil sample (charge 1), while Figure 6.2 shows the analysis of one green oil sample (terpene components, except γ-terpinene and terpinolene, are not marked). The richest Brazilian oil profile can be then visually well observed.

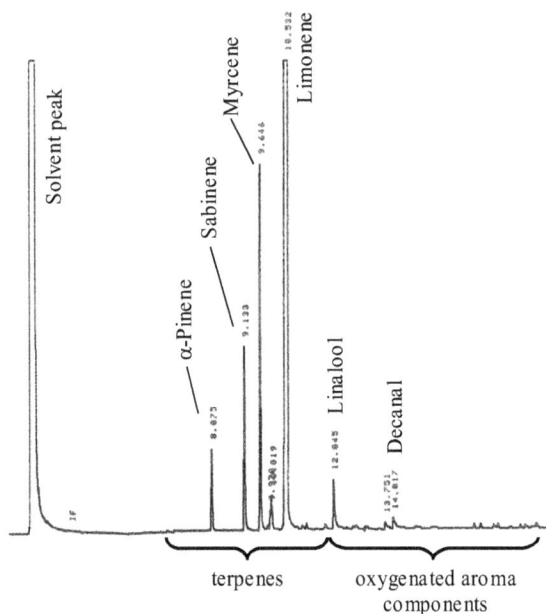

Figure 6.1. Gas-chromatograms obtained for the raw Spanish red oil (charge 1).

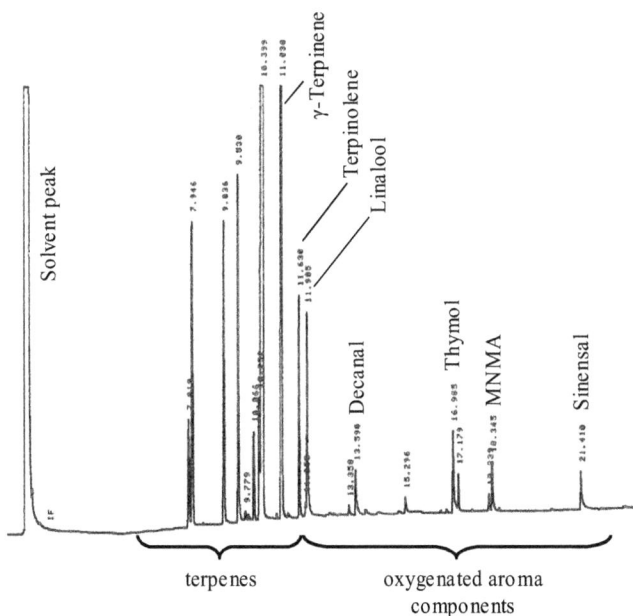

Figure 6.2. Gas-chromatograms obtained for the raw Brazilian green oil.

6.2. Phase Equilibria

VLE data of binary systems comprising CO_2 and a pure oil component or CO_2 and the multicomponent mixture (raw citrus oil) have been extensively investigated, mostly because citrus oil deterpenation was the main goal of these studies. VLE data of ternary systems have also been investigated, since they are interesting for deterpenation processes, providing the knowledge on the phase behavior of some volatile aroma components that must be separated from the most abundant terpenes. Table 6.2 summarizes some works on VLE experimental data for different citrus oil systems, including binary and ternary ones.

Table 6.2. VLE data for diverse citrus oil systems.

System	T(°C) / P(MPa)	References
CO_2+limonene	34-76 / 7.7-14.1	[146]
	40-60 / 5.1-12.6	[147]
	40-60 / 1.0-11.0	[148]
	50-70 / 5.1-12.6	[149]
	50-70 / 5.1-10.6	[150]
CO_2+ linalool	40-60 / 4.0-11.0	[151]
	50 / 7.5-8.4	[150]
CO_2+lemon oil	40-70 / max. 19.0	[152]
	40-70 / 7.0-8.5	[153]
CO_2+bergamot peel oil	40-70 / max. 19.0	[152]
CO_2+orange peel oil	40-70 / 8.3-12.4	[154]
	50-70 / 8.0-13.0	[112]
	50-70 / 7.0-13.5	[155]
CO_2+limonene+linalool	40 / 7.5-8.9	[150]
	42 / 6.3-8.7	[149]
	20-76 / 5.0-14.0	[156]

The large number of publications for the system CO_2+limonene demonstrates the importance of evaluating its phase behavior, since mutual solubilities and critical pressures of CO_2+raw citrus oils can be estimated directly from CO_2+limonene VLE data, especially for red MPO and other oils with a limonene content of 95 wt.-%, as reported by Budich [112].

Studies on pseudo-binary systems composed by CO$_2$ and natural complex mixtures (raw citrus oils) usually require sophisticated analysis methods. Conventional analysis are normally not able to detect and identify correctly all components present in the chosen samples, especially when dealing with isomers and high molecular weight components. As mentioned in the literature [112], other problem is related to a possible change in the oil compositions when using different raw material charges (see Table 6.1). Additionally, the oil origin and previous extraction methods can deeply affect its composition profile.

Ternary mixtures composed by CO$_2$+limonene+one representative aroma component can be treated as model mixtures for VLE measurements of raw citrus oils. The phase behavior of this ternary mixture can then represent with good accuracy the behavior of a real multicomponent mixture. That was observed when comparing the behavior of the ternary systems CO$_2$+limonene+linalool with the binary CO$_2$+orange peel oil [112, 156].

In order to investigate a wide range of conditions, ethane has been suggested as a possible alternative to replace CO$_2$ for the supercritical deterpenation of citrus oils. VLE data of binary systems composed by ethane+limonene and ethane+linalool have been published in the last years [157, 158], as well as ternary systems composed by ethane+limonene+linalool [159], which can represent with good accuracy the pseudo-binary system ethane+orange peel oil [160].

6.2.1. VLE Data

Based on the information provided above, the VLE data previously published by Budich [112] for orange peel oil will be used in this work. Most measurements were performed at 50, 60 and 70 °C from 8 MPa to pressures close to the one-phase region. Figure 6.3 illustrates the mutual solubility data obtained for the system CO$_2$+orange peel oil.

The solubility of orange peel oil components in SC-CO$_2$ was low below 8 MPa for the studied conditions. This could be expected, since the same behavior was observed for the binary CO$_2$+limonene. Smaller mutual solubilities were obtained at 50 °C at pressures lower than 8 MPa. The tendency observed at isothermal conditions showed an increase in the mutual solubility with increasing pressure until the system was completely mixed at the critical point.

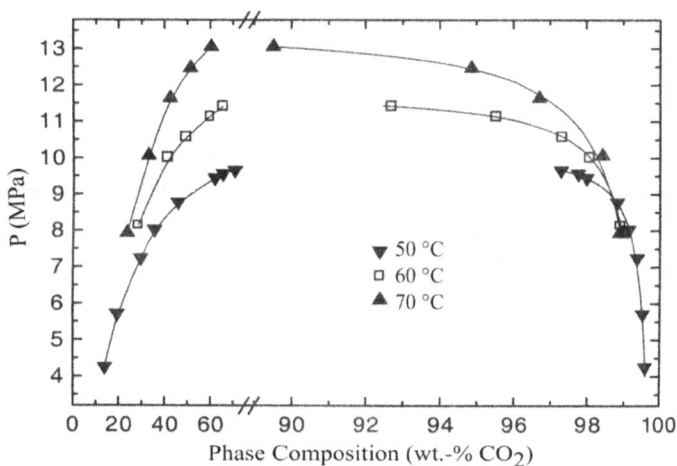

Figure 6.3. VLE data for the system CO_2-orange peel oil [112].

Since the percentage of limonene exceeds 95 wt.-% in the feed mixture, small differences in the total concentrations of either terpene or oxygenated aroma components have no influence on the mutual solubility. This was observed by Budich [112] with three different oil samples, with overall compositions varying from 98.25 to 98.77 wt.-% of total terpenes and from 1.23 to 1.75 wt.-% of total oxygenated aroma components. This consideration was also applied in this work, because the total red MPO terpene concentrations varied from 98.5 to 99.2 wt.-% (Table 6.1).

Budich have also compared the quasi-binary data for different orange oil mixtures. VLE data of a terpene rich fraction (0.15 wt.-% aroma components) and a 5-fold concentrate fraction (8.01 wt.-% aroma components) were determined at the same temperatures discussed above. The influence of composition on mutual solubility is presented in Figure 6.4. As can be observed, mutual solubility data for CO_2+orange peel oil lied between both fractions, presenting smaller deviations in comparison with the enriched terpene fraction (CO_2+terpenes, 99.85 wt.-%). It was also clear that the mutual solubility decreased with the increase of high-boiling (high molecular weight) aroma components in the liquid phase, such as waxes, which ones are practically insoluble in CO_2. A further investigation of the influence of product composition on mutual solubility is necessary, but is beyond the scope of this work.

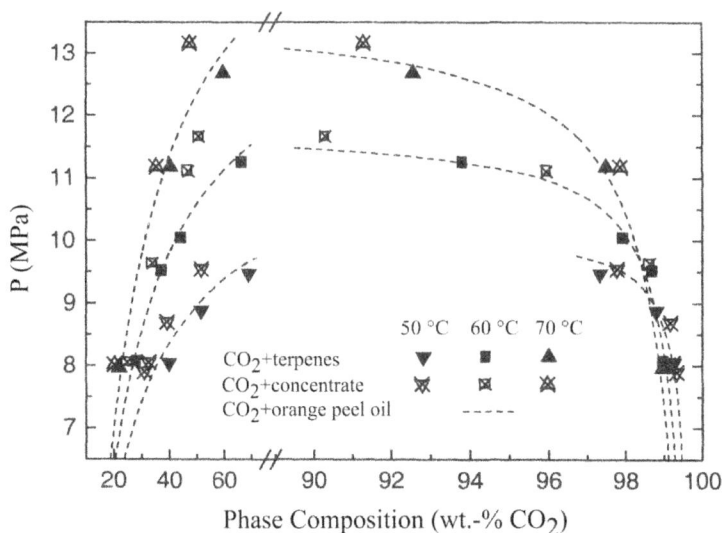

Figure 6.4. Influence of product composition on mutual solubility [112].

Figure 6.5 illustrates the separation factor profiles as a function of pressure at three different isotherms for the mixture CO_2+orange peel oil [112]. It is mentioned that, because of the different raw material charges used, separation factors were not the same. Indeed, the sampling procedures were also presented by the author as a source of deviations in the measurements. Therefore, obtained data were fitted and appropriate equations were proposed. The dashed lines in Figure 6.5 represent the calculated values.

Mention to the relationship between loading (g extract/kg CO_2) and selectivity must also be done. The optimal conditions for a separation task are achieved when high loading is combined with a high selectivity. Since the process is conducted under isothermal conditions, separation factors tend to decrease at higher loadings. When the critical point of the mixture is reached, no separation is possible (homogeneous mixture) and the separator factor turns equal to 1. The tendency of the dashed lines at near-critical conditions was in good agreement with critical values for the binary system CO_2+limonene, what suggested once more that VLE data for the multicomponent system CO_2+orange peel oil can be estimated from the equilibrium data available for this pseudo-binary system.

The temperature influence on vapor pressures of limonene/linalool and limonene/decanal is presented in Figure 6.6 [161]. As can be observed, changes in vapor pressure are larger for the oxygenated components when temperature increases. The separation factor between terpenes and aroma must decrease if the relative concentration of oxygenated components in the vapor phase increases.

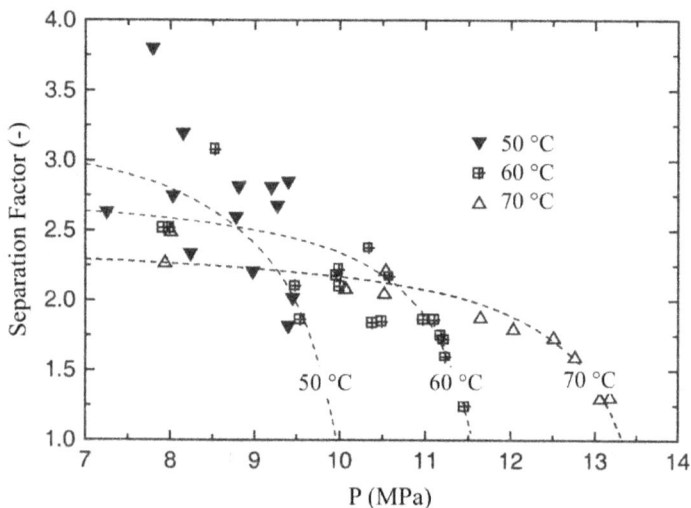

Figure 6.5. Separation factors between terpenes and aroma [112].

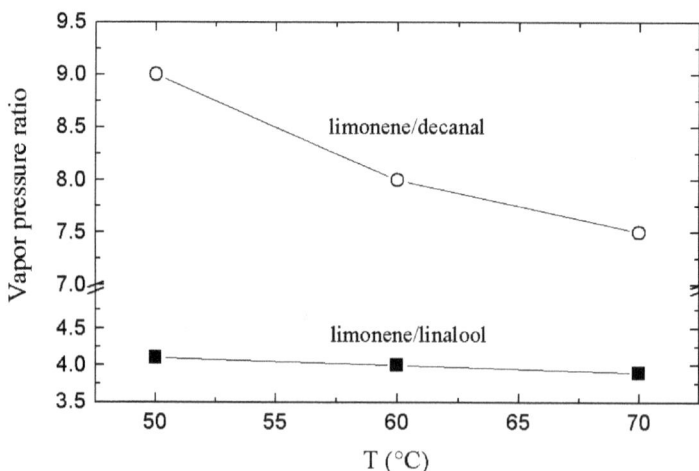

Figure 6.6. Temperature effect on the vapor pressures of pure aroma components [161].

VLE measurements were also performed for samples obtained from countercurrent experiments in order to determine the composition dependency of the separation factor. Samples rich in terpenes (extract) and aroma components (raffinate) were used. Results obtained at 60 °C and 10 MPa are illustrated by triangular diagrams in Figures 6.7 and 6.8 [112]. The terpene concentration affected strongly the phase behavior of the mixture, since their concentrations were always between 95 and 96 wt.-%. Higher terpene concentrations in the liquid phase led to an increase in the content of CO_2 in the liquid phase, as well as terpene solubilities in the solvent.

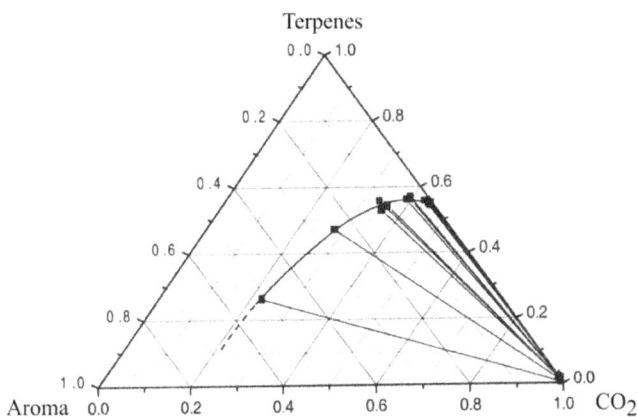

Figure 6.7. Gibb´s diagram representation for the system CO_2+terpenes+aromas [112].

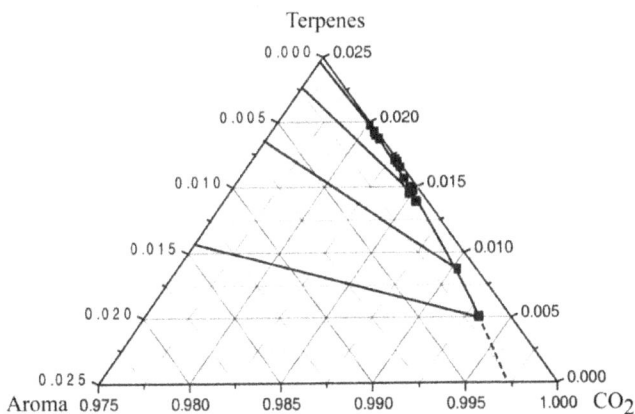

Figure 6.8. Right corner of the triangular diagram presented in Figure 6.7 [112].

In order to provide a better understanding of the phase behavior of the system, a different representation should be employed, since the influence of the solvent must be considered for the stage calculations. VLE data were then better represented by a Jänecke diagram, as presented in Figure 6.9 [112]. When combining the data shown in Figures 6.9 and 6.10, the determination of the number of theoretical stages for this determined system at 60 °C and 10 MPa could be performed. By observing Figure 6.9, a linear decrease of the solvent ratio in the vapor phase line corresponded to an increase in the total terpene content. Below 70 wt.-% terpenes, the solvent solubility in the liquid phase remained practically constant. However, it changed drastically at higher terpene concentrations, especially at concentrations higher than 95 wt.-%.

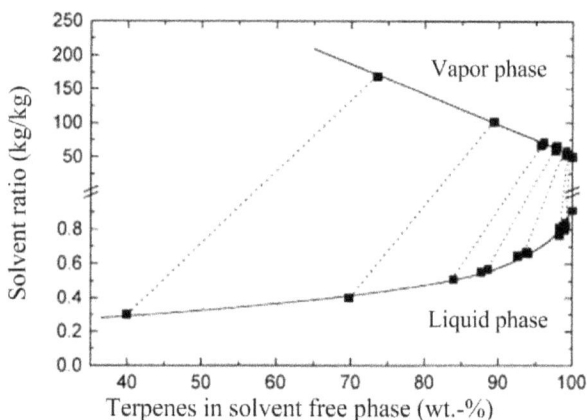

Figure 6.9. Jänecke diagram for the system CO_2+orange peel oil (60 °C, 10 MPa) [112].

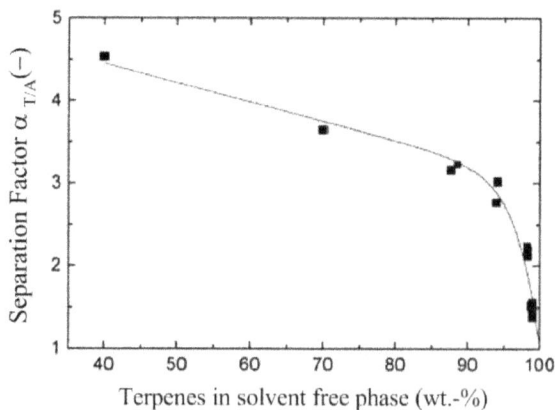

Figure 6.10. Separation factors at different liquid compositions (60 °C, 10 MPa) [112].

From Figure 6.10 it can be observed that separation factors decreased when increasing the total terpene content in the liquid phase. Additionally, good solvent selectivities can be obtained at low terpene concentrations. At terpene concentrations higher than 95 wt.-%, the smooth linear decrease changes to a very significant decrease trend.

According to Budich [112], the changes in the separation factors at high terpene concentrations are related to the complex composition of the so-called aroma fraction. When paying attention to separation factors of terpenes and one single oxygenated component, such as linalool, it can be observed that its value remained practically constant at 60 °C and 10 MPa, in the range of 1.2. For the mixture terpenes+decanal at the same conditions, this value lied between 1.2 and 1.4, and for the total aroma fraction it varied from 2.2 to 1.2. Thus, the minimum value to be mentioned as acceptable for the stage calculations was the one obtained for terpenes+linalool (1.2), since linalool presents the highest affinity to limonene in comparison with other aroma components. This was considered by Budich in the stage calculations for the system CO_2+orange peel oil and will also be used in this work for the system CO_2+MPO, as presented in the next Chapters.

6.3. Countercurrent Fractionation Experiments

In this Chapter, the results obtained for the countercurrent fractionation of MPO using SC-CO_2 as solvent are presented. Experiments were conducted on the 7 m high column described in Chapter 5.2.3. Sets of experiments without extract reflux were only performed for the Spanish red MPO, whereas experiments with extract reflux were carried out for both raw materials investigated at temperatures varying from 50 to 70 °C and pressures between 8.0 and 11.5 MPa.

The results presented in the following Chapters were obtained after the steady state condition was achieved. For the systems investigated, steady state was achieved approximately 75 minutes after starting the experiment, in average. Sequential samples were then analysed through GC immediately after sampling (see Chapter 5.1.2). Based on GC analysis, some samples (especially raffinate ones) were analysed through GC-MS in order to identify a larger number of components, because some of them could not be detected by the conventional GC analysis.

6.3.1. Experiments without Extract Reflux

The results obtained through countercurrent extraction experiments of the red Spanish oil without reflux are presented in this Chapter. The evaluated operational parameters were pressure, temperature, solvent flow and feed flow rates. Solvent density and solvent-to-feed ratio (SFR) were directly calculated from the operational parameters, as presented in Table 6.3. The relation between feed and product, represented by the raffinate flow, was described by the folding ratio (FR). By performing the experiments without extract reflux, the obtained folding ratios were very low, what would not allow a sufficient aroma concentration in the raffinate samples.

Table 6.3. Countercurrent extraction data for the Spanish red oil without reflux.

P (MPa)	T (°C)	ρ_{CO_2} (kg/m³)	CO_2 (kg/h)	Feed (g/h)	SFR (kg/kg)	FR (kg/kg)
8.0	50	220	2.5	145	17.24	1.08
8.5	50	250	3.0	58	51.72	1.16
9.0	50	286	2.7	65	41.54	2.00
9.5	50	331	2.6	66	39.39	2.20
9.5	70	227	2.0	125	16.00	1.12
10.0	70	248	2.0	120	16.67	1.20
11.0	70	294	1.8	126	14.29	1.34
11.5	70	320	1.9	130	14.62	1.82

Increasing the folding ratio by increasing CO_2 density and flow rate was limited by the hydrodynamic behaviour of the two phases system in the column. If the flow rate was kept high, the column would be susceptible to flooding. Increasing the density of the vapor phase and thus decreasing the density difference between the phases, would also favor flooding. Therefore, operation close to the flooding point made best use of the column capacity [112].

The separation performance was expressed by the overall selectivity for the key-components limonene and linalool ($\beta_{lim/lin}$). The selectivities were calculated as the quotient of the component concentration ratios between extract and raffinate samples, as presented by Equation 6.1:

$$\beta_{i/j} = \frac{y_i^{Ex} / x_i^{Raf}}{y_j^{Ex} / x_j^{Raf}} \qquad\qquad (6.1)$$

where *Ex* refers to the extract sample and *Raf* refers to the raffinate fraction.

Additionally, overall selectivities for the pseudo-components terpene and aroma fraction ($\beta_{T/A}$) were determined. Figure 6.11 shows the overall selectivities obtained at 50 and 70 °C in dependence of CO_2 density. There was a significant difference between the selectivities $\beta_{T/A}$ and $\beta_{lim/lin}$ for the performed experiments. That could be explained by the low differences in the solubilities between limonene and linalool at these determined conditions. In all cases, the extracts were free from aroma components despite linalool, while limonene was the main component of the terpene fraction in extracts and raffinates. At the achieved folding ratios and selectivities, the aroma losses in the extract fraction was between 2.5 and 17 wt.-%, depending mainly on the folding ratio value, since linalool concentrations were not affected by these variations. This behavior was also observed by Budich [112] in the deterpenation of orange peel oil. For the calculation of the selectivities, not all the most representative aroma components could be detected in the extract samples, since their concentrations were very low. However, their enrichment was observed in the raffinate, especially when performing experiments with extract reflux (see following Chapter).

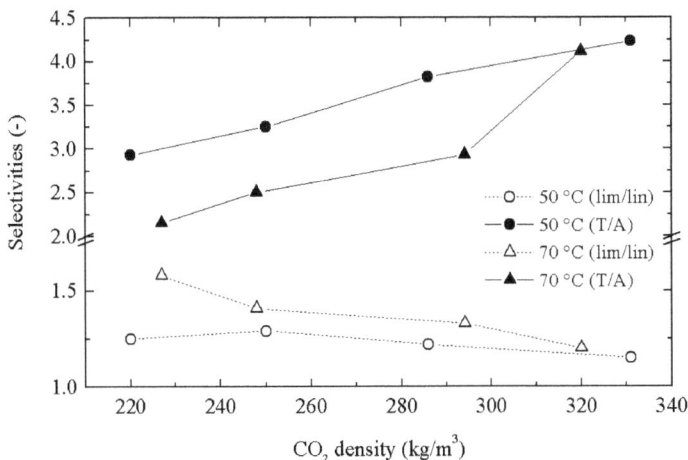

Figure 6.11. Overall selectivities $\beta_{T/A}$ and $\beta_{lim/lin}$ obtained at 50 and 70 °C.

6.3.2. Experiments with Reflux using Red Oil

By introducing extract reflux, higher extract purities are intended to be achieved. The concentrations of aroma components in the extract must be minimized, while the folding ratio should be maximal, providing then higher enrichments of aroma components in the raffinate.

The influence of the extract reflux was investigated at 50 °C and at different pressures. The results obtained are presented in Figure 6.12 as a function of the loadings used (g of extract/kg of solvent) and were following the trends obtained by Budich [112] at 60 °C in the deterpenation of orange peel oil.

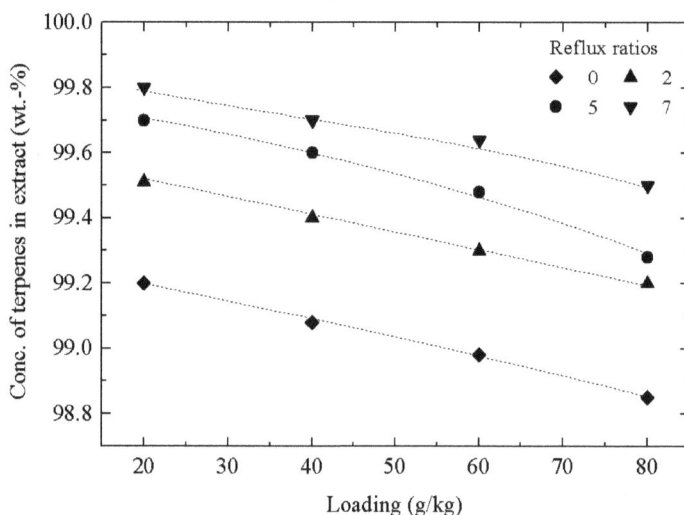

Figure 6.12. Influence of the reflux ratio in the fractionation of red MPO at 50 °C.

As observed by Budich [112], high terpene concentrations were obtained at low loadings, in the pressure range between 9 and 10 MPa. Extract compositions were slightly higher when increasing reflux ratios, especially from 5 to 7. Reflux ratios higher than 7 were not investigated due to equipment limitations. Minimum extract concentrations should be equal to the total terpene concentration of the feed material. In this case, it corresponded to 98.52 wt.-% (red oil charge 2). This value should be reached at higher loadings. At higher loadings, the concentrations in the extract decreased due to the decrease in the selectivities.

Table 6.4 shows experimental data for the countercurrent fractionation of the Spanish red oil. When necessary, the extract flow and the folding and reflux ratios were corrected, considering the feed flow as the sum of extract and raffinate. This was attempted in order to compensate the losses found in the mass balances, what was probably due to the packings hold-up. Once more, all extract samples after starting the experiment were free of quantitatively detectable oxygenated compounds, exception made only for linalool.

Table 6.4. Experimental data for countercurrent extraction of red oil with extract reflux.

P (MPa)	T (°C)	ρCO_2 (kg/m³)	CO_2 (kg/h)	Feed (g/h)	SFR (kg/kg)	FR (kg/kg)	$\beta_{lim/lin}$ (-)	$\beta_{T/A}$ (-)	Reflux (g/h)
8.5	50	250	3.2	95	34	1.29	1.48	3.58	18.0
8.7	50	263	4.0	100	40	1.25	2.09	5.34	49.4
9.0	50	286	3.9	98	40	2.54	1.41	3.52	74.6
9.5	50	331	3.7	45	82	2.50	1.50	3.70	52.0
8.5	60	212	2.5	48	52	3.05	3.20	3.57	55.0
9.0	60	235	2.7	44	61	3.94	2.98	4.25	60.0
9.5	60	261	2.8	50	56	4.24	2.34	5.02	58.0
10.0	60	291	2.5	52	48	5.10	2.52	5.26	57.5

Overall selectivities at the chosen process conditions were in the same range or higher than those obtained without reflux (see Chapter 6.3.1). Comparing experiments carried out at 8.5 MPa and 50 °C with and without reflux (Figure 6.11), additional extract reflux was not able to improve the overall selectivities. This was then assumed to be related to the low aroma concentration in the feed samples, which ranged between 0.37 and 0.54 wt.-% (see Table 6.1).

In order to simulate a higher column with more separation stages, raffinate samples obtained at 9.0 MPa and 60 °C were used as feed material. This experiment showed high selectivities and the results obtained are presented in Table 6.5. Calculated selectivities suggest that, as expected, a higher column could improve separation performance.

A concentration dependence of the separation factor for orange peel oil was investigated by Budich [112]. The developed model showed a decrease of the separation factor at terpene concentrations above 98.25 wt.-%, what can be explained by the decrease of less volatile components among the non-terpenes with increasing terpene content. As soon as a low volatile substance is not present anymore in a mixture, its comparatively high separation factor will not contribute to the overall total separation factor.

According to Budich [112], the separation factors with respect to a single substance do not change significantly in this concentration range. Thus, the separation factor of linalool, which is the most soluble aroma component in CO_2, was suggested as a lower limit. In order to verify the influence of linalool concentration on the separation performance, one experiment was carried out with a feed enriched to 4 wt.-% linalool at 10 MPa and 60 °C (see Table 6.5). The selectivity obtained for limonene/linalool separation was 6.6, a value higher than the ones obtained at the same conditions with crude MPO and raffinate being used as feed material. Due to the high linalool concentration in the aroma fraction, the other aroma components were not able to contribute significantly to the selectivity $\beta_{T/A}$. As a result, it was only slightly higher than $\beta_{lim/lin}$. An increased aroma content would improve the separation, as results obtained by several authors with model mixtures have shown [137, 162].

Table 6.5. Experimental data obtained for red MPO with different feed samples.

Feed samples	CO_2 (kg/h)	Feed (g/h)	SFR (kg/kg)	FR (kg/kg)	$\beta_{lim/lin}$ (-)	$\beta_{T/A}$ (-)
Crude MPO	2.2	43	51	1.51	1.52	3.26
Raffinate	2.3	41	56	1.87	2.89	9.19
Linalool enriched MPO	2.1	41	52	1.21	6.60	7.17

6.3.3. Experiments with Reflux using Green Oil

When performing the countercurrent fractionation of Brazilian green MPO, the overall selectivity was evaluated through the separation analysis of the key-components terpinolene and linalool, since their separation was more difficult to be achieved. Overall selectivities for terpene and aroma fraction were also calculated and the obtained results are presented in Table 6.6. In some extract samples, the aroma concentration could not be measured quantitatively because it was below the GC detection limit, what proved that the complete elimination of terpenes was achieved. Therefore, the calculated selectivities were based on assumed aroma concentrations of 0.08 wt.-%, what was considered a reliable value according to Donoso [163].

Table 6.6. Experimental data for countercurrent extraction of Brazilian oil with reflux.

P (MPa)	T (°C)	ρ_{CO_2} (kg/m³)	CO_2 (kg/h)	Feed (g/h)	SFR (kg/kg)	FR (kg/kg)	$\beta_{tpl/lin}$ (-)	$\beta_{T/A}$ (-)	Reflux (g/h)	RR (-)
8.5	50	286	3.0	69	43	4.20	1.11	7.22	49.4	1.0
9.0	50	286	3.1	71	52	4.45	1.29	10.27	84.0	1.5
10.0	50	386	3.0	74	40	5.21	3.45	11.24	85.0	2.0
8.5	60	212	2.2	100	22	4.02	1.98	8.24	100.0	3.0
9.0	60	235	2.0	102	19	4.15	2.54	10.89	111.0	3.1
10.0	60	291	2.1	104	20	5.03	4.81	12.80	110.1	3.9
Feed enriched to 1.1 wt.-% linalool										
9.0	50	286	3.0	73	49	3.79	1.30	6.37	79.4	1.5
10.0	60	291	2.9	78	37	4.22	4.35	8.34	105.1	2.0

By comparing the results presented in Table 6.4 with the ones presented above, it can be seen the tendency of higher linalool concentrations in the aroma fractions (raffinate) and lower concentrations in the extract when using green oil. The higher reflux ratio and the lower linalool concentration in the feed contributed to this. The difference in overall selectivity with respect to terpenes and aroma is even larger due to the presence of less volatile aroma components in the Brazilian green oil. The effect of linalool concentration became apparent when comparing the last experiments presented in Table 6.6. In analogy to the procedure adopted for the red oil experiments (Table 6.5), the last two experiments were conducted with increased linalool content, which varied from 0.14 to approximately 1.1 wt.-%. All other process conditions were similar. The resulting selectivities with respect to terpinolene and linalool followed the same trend when using the original Brazilian oil, while selectivities between terpene and aroma fraction decreased. That showed that the linalool separation factor did not change significantly in this concentration range and was lower than for other aroma components. The increase in the linalool concentration has caused the decrease in the separation factor between aroma and terpene components.

6.3.4. Evaluation of the Separation

After performing the countercurrent experiments, the next step corresponded to the evaluation of the process, in order to perform the column design for a reliable separation task.

As mentioned in the previous Chapters, high selectivities were obtained for the set of experiments carried out. The selectivities with respect to terpene and aroma fraction obtained with the Brazilian green oil were generally higher than the ones obtained with the Spanish red oil due to the higher aroma concentration in the feed. By increasing linalool concentrations in the feed material up to 4 wt.-%, an increase in the separator factor and, consequently in the selectivity, was obtained. At higher linalool concentrations, higher separation factors were achieved, as observed by other authors [112, 162]. Additionally, small changes in linalool concentrations were not able to change separation factors significantly.

6.3.4.1. Determination of the Number of Theoretical Stages

As presented in Chapter 2.3, the number of stages required for a determined separation task can be determined by using the method of Ponchon-Savarit plotted in a Jänecke diagram. When using this method, the amount of solvent is usually expressed as solvent ratio and can be plotted in function of a pseudo-component concentration in the solvent-free liquid phase.

In this work, the phase envelope and tie lines were evaluated from previously measured phase-equilibrium data, as presented in Chapter 6.2.1. The extract and raffinate lines, as well as separation factors, were obtained from correlations determined by Budich [112] for the system orange peel oil.

Beyond 85 wt.-% terpene content in the solvent-free liquid phase both lines were assumed to be straight lines. The CO_2 content at $x_T = 98.25$ %, which was derived from binary VLE-correlations, with an assumed solvent ratio of 0.5 at $x_T = 85$ wt.-% were used to calculate the equation for the raffinate line.

The resulting equations used for the raffinate lines were (Equations 6.2 and 6.3):

$$S / F_{liq} = 0.0396 \cdot x_T - 2.8627 \text{ at 10 MPa, 60 °C} \tag{6.2}$$

$$S / F_{liq} = 0.0392 \cdot x_T - 2.8360 \text{ at 9 MPa, 50 °C} \tag{6.3}$$

Since the vapor phase curves in the Jänecke diagram were supposed to be straight lines, the slope of these line should be obtained. The slope of the lines was then obtained by Budich [112] through derivations from own experimental data and literature data for CO_2 and limonene, the most representative terpene component in citrus oils. The following correlations were then used for the determination of the slope of the extract line (Equations 6.4 and 6.5):

$$Slope_{Ex} = 12.7 \cdot P - 125 \quad \text{at 50 °C} \tag{6.4}$$

$$Slope_{Ex} = 2.9 \cdot P - 33.9 \quad \text{at 60 °C} \tag{6.5}$$

The vapor-phase concentration in equilibrium with $x_T = 98.25$ wt.-% and the solvent-ratio calculated by the VLE-correlation for the vapor phase defined a point on the extract line with the coordinates $y_{T,vap} = 99.245$ wt.-%, $S/F_{vap} = 51.561$ at 10 MPa and 60 °C. This point and the slopes calculated by Equations 6.4 and 6.5 yielded the equations for the extract lines (Equations 6.6 and 6.7):

$$S / F_{liq} = -4.9 \cdot x_T + 538 \quad \text{at 10 MPa, 60 °C} \tag{6.6}$$

$$S / F_{liq} = -10.7 \cdot x_T + 1103 \quad \text{at 9 MPa, 50 °C} \tag{6.7}$$

Separation factors showing the relations between the terpene concentrations in vapor and liquid phases were determined by means of two different models developed by Budich [112]. By using ternary VLE-data, a correlation at 98.25 wt.-% overall terpene content of the oil was obtained. The first model presented a constant separation factor up to $x_T = 98.25$ wt.-% and a linear declination to 1.1 at 100 wt.-% afterwards (Equations 6.4 and 6.5).

On the other hand, an improved model was proposed in order to account for the non-linearity of the separation factor decrease at higher terpene concentrations. This model described better the separation factor in the range of $x_T = 98$ to 100 wt.-%, with an equation following the form presented by Equation 6.8, achieving a minimum separation factor value of 1.05 at 100 wt.-%:

$$\alpha_{T/A} = \frac{1}{A + B \cdot x_T} \tag{6.8}$$

Further discussions on the determination of the constants A and B can be found in the work of Budich [112]. The separation factors obtained according to the mentioned models at 9 and 10 MPa and at 50 and 60 °C are presented in Figure 6.13. The differences between both models can be perfectly visualized at high terpene concentrations.

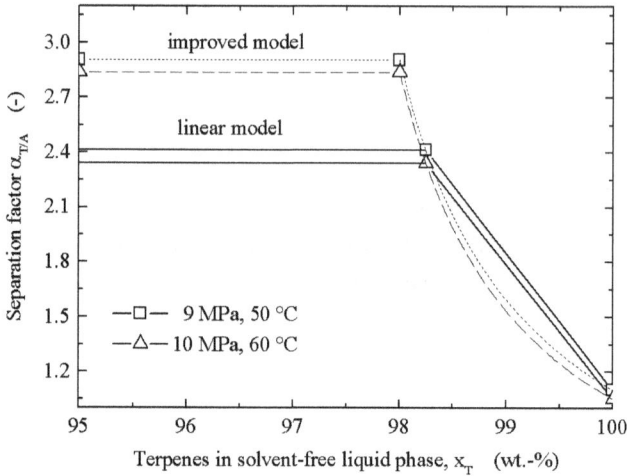

Figure 6.13. Comparison between both models at different experimental conditions.

In order to perform the design of a multistage column for the fractionation of MPO, a Jänecke diagram for one experiment carried out at 10 MPa and 60 °C was obtained, as can be seen in Figure 6.14.

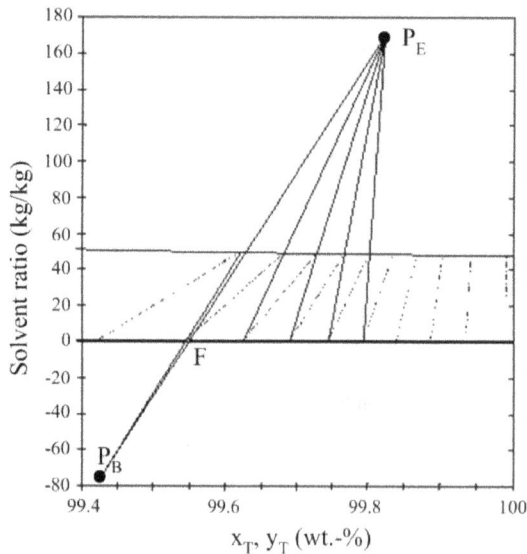

Figure 6.14. Jänecke diagram for one experiment with red oil at 10 MPa and 60 °C.

Figure 6.14 shows that one or two stages were determined for the stripping section and 5 stages for the enriching section by using the linear model for the separation factors. As expected, this depended strictly on the chosen model for the separation factor, since the improved model predicted 5 stages for the stripping section and 39 for the enriching section. Both models were based on data measured at terpene concentrations of 99 wt.-% and one assumption for 100 wt.-%. Therefore, no measurements existed in the concentration range considered.

The height equivalent to one theoretical stage (HETS) was then determined as 3/1.5 m for the enrichment and 0.6 m for the stripping sections when using the linear model, whereas these values changed to around 0.6 m for the enrichment and 0.07 m for the stripping sections when using the improved model. As mentioned by Budich [112], near the flooding point, backmixing of the phases reduces the HETS values, what can explain the smaller HETS obtained when using the improved model.

In the case analysed, parameters and measurements of the experiment fitted well to the values predicted through mass and component balances, which ones are shown in the Jänecke diagram. However, reflux ratio was too high. The reflux ratio can be read from the diagram by dividing the vertical distance from P_E to the extract line by the vertical distance from the abscissa at x_E to the extract line. The value read from the diagram in this case was 2.54, while the measured reflux flow and the extract flow yielded a reflux ratio of only 1.51.

Furthermore, additional experiments showed differences in the component mass balances that persisted even when correcting the overall mass balance. The problem with these calculations consisted on the fact that the concentrations were mainly determined by the very small concentrations of linalool compared to the terpene ones. Therefore, the deterpenation of MPO by means of ad-/desorption in silica gel with SC-CO_2 was suggested as a process to be coupled with countercurrent fractionation, aiming a better separation performance between terpene and aroma components, what will be discussed in the following Chapters.

6.4. Selective Ad-/Desorption Fractionation

The deterpenation of MPO must be carried out in order to separate the undesired terpene components from the oxygenated aroma fraction. Besides countercurrent extraction with SC-CO_2, other good alternative is the selective ad-/desorption fractionation. By adsorbing different MPO samples in silica gel, a sequential selective desorption process using CO_2 as solvent at different pressure levels can be employed. Silica gel 60 (Fluka, Switzerland) was used as adsorbent for these experiments; its average pore size was 60 Å and the particle size was 0.2-0.5 mm.

Once the adsorption was completed, the unpolar terpene components were then firstly desorbed when applying lower operational pressures (around 8 MPa), as presented in Chapter 4.3.1. After the desorption of terpenes was achieved, the system pressure was increased (up to 20 MPa) in order to desorb the highly desired polar aroma components, which remained adsorbed at lower pressures. In this work, selective desorption experiments were conducted with crude Spanish red and Brazilian green oils. Additionally, raffinate samples of the Spanish red oil obtained through the previous countercurrent experiments were also used as feed material.

6.4.1. Preliminary Experiments

Preliminary experiments conducted by Schwänke [164] using a 100 ml extractor determined feasible conditions for the selective desorption of MPO. First of all, the influence of the CO_2 flow rate on the oil recovery was studied. The results obtained are presented in Table 6.7.

The determination of the time necessary for the deterpenation at lower pressure (8 MPa), i. e., before the pressure step, was chosen based on the amount of solvent passed through the extractor, which should be the same for the different solvent flow rates used. As shown in Table 6.7, the recovery, i. e., the percentage of the feed collected when sampling, was only 46, 9, and 6 wt.-%. These values were considered insuficient because the losses were very high. As observed by other authors [123], high-volatile monoterpenes can be lost with the CO_2-effluent. To avoid terpene losses, the sampling vials downstream the expansion valve must be cooled with ice or the solvent flow rates must be lower.

Table 6.7. Results obtained for the preliminary selective desorption experiments.

Flow rate (kg/h)	time at 8 MPa (min)	time at 20 MPa (min)	recovery (wt.-%)
0.8	90	60	46
1.2	60	45	9
1.6	45	120	6

Based on this assumption, the following experiments showed higher recoveries and an additional cooling was used for all preliminary MPO desorption experiments. The yield in mass percent (g sample/g feed) of the originally adsorbed Spanish oil raffinate is presented in Figure 6.15. The amount of oil adsorbed corresponded to 10 wt.-% of the silica gel mass used, which was 30 g. The pressure steps are indicated by the dashed lines. Both experiments showed that a better separation with selective desorption was possible. GC-analysis of the samples with sufficient amounts of extract taken after the pressure step showed limonene concentrations lower than 10 wt.-%.

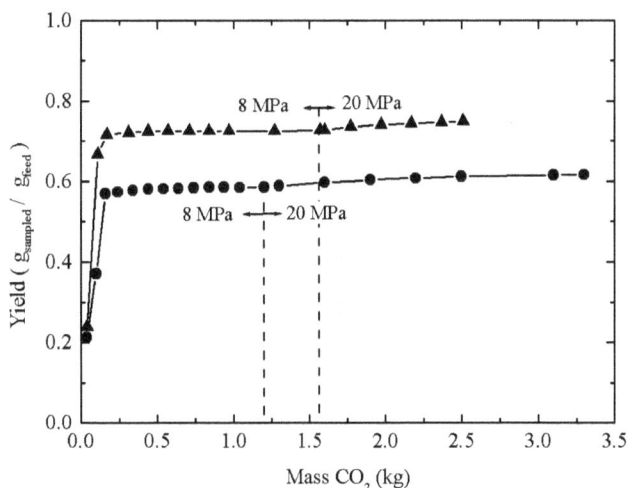

Figure 6.15. Selective desorption at 40°C and 10 wt.-% loading at different CO_2 flow rates: 0.8 kg/h (triangles) and 1.2 kg/h (circles).

6.4.2. Selective Desorption – Experimental Results

6.4.2.1. Experiments with Spanish Oil Raffinate Samples

Based on the preliminary experiments, the sequential step was related to the verification of the optimal experimental conditions for the systems evaluated in this work.

Experiments using Spanish oil raffinate samples as feed and employing different solvent flow rates were carried out. Flow rates of 2 and 4 l/min were found to be the operational conditions investigated, especially because these flow rates were restricted by technical limitations of the apparatus. 20 g of adsorbent were used with a loading of 10 wt.-% raffinate sample. At standard conditions, these values corresponded to 0.23 and 0.47 kg CO_2/h, respectively. The solvent-to-adsorbent ratio ranged from 10.9 – 21.8 kg CO_2/(h kg silica), whereas for the preliminary studies were in the range of 26.7 – 53.3 kg CO_2/(h kg silica).

For the experiments performed in the next Chapters, no additional cooling was necessary. In order to minimize the Joule-Thompson effect, the expansion valve heating was adjusted to allow the complete solute-solvent separation without blocking the valve, minimizing additionally eventual extract losses. Figure 6.16 shows the extraction curves obtained at 40 °C.

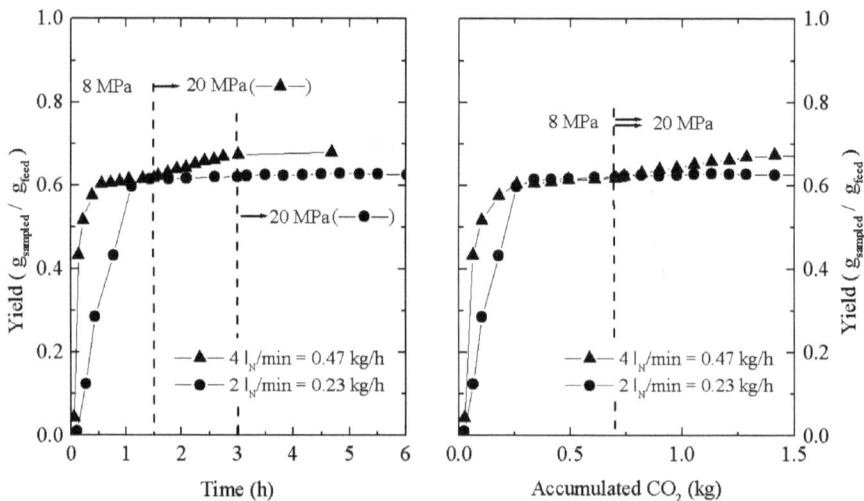

Figure 6.16. Percentage of raffinate samples recovered in dependence of CO_2 mass.

When comparing the overall extraction curves, it can be observed that the amount of obtained extract depended mostly on the total amount of solvent rather than its flow rate. As expected and as mentioned in several textbooks [1, 2], extraction at a higher flow rate achieved a higher yield at a shorter operational time. Both curves reached a fixed plateau after the same amount of solvent flowed through the extractor. However, the initial slope of the curves were different: after 30 minutes, the yield (g oil/g feed) obtained for the higher flow rate was approx. 60 % of the initial value, while for the smaller flow rate it achieved not more than 30 %. Then, an increase of the flow rate could be advantageous. This increase in the initial extraction rate with increasing flow was observed by Rodrigues et al. [165] and is related directly to the decrease in axial dispersion.

In order to evaluate the deviations in the process, assays with raffinate and crude Spanish oil under identical conditions were performed. By observing Figure 6.17, it can be observed that a loading of 25 % provided a total oil recovery of more than 85 %. The diffusion controlled phase was reached when the amount of CO_2 consumed was approx. 0.35 kg. The extraction may be described by an equilibrium model if the percentage of desorbed components depends only on the amount of solvent used, what occurs in the desorption of limonene and linalool. This has been already observed in the deterpenation of bergamot oil and citrus oil model mixtures [123, 166]. Based on the experiments obtained, the operational pressure must be increased after 0.41 kg of CO_2 flowed through the extractor. For all following experiments a flow rate of 3.5 l CO_2/min (0.41 kg/h) was used.

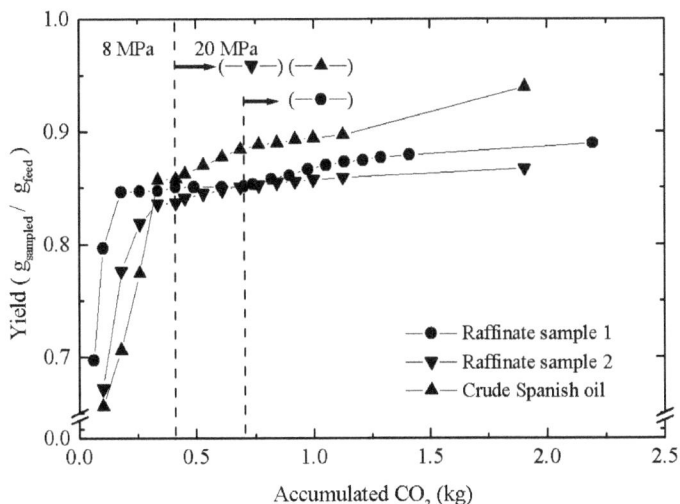

Figure 6.17. Yield of desorption of raffinate and crude red oil at 40 °C, loading 25 %.

6.4.2.2. Experiments with Red Spanish Oil

The following experiments with original red Spanish MPO have been performed after the sets of preliminary experiments and experiments with raffinate from the countercurrent investigations have provided sufficient operational information about the process.

Investigations on the best operational conditions, namely temperature, pressure and loading, have been performed in order to achieve the complete deterpenation of the oil samples. The ideal conditions would achieve an enrichment of oxygenated aroma components at mild temperatures and pressures, using a reliable amount of silica gel at relatively short extraction times.

6.4.2.2.1. Effect of Loading

The influence of loading on MPO desorption was investigated at 40 °C and with pressures of 8 and 20 MPa. Different loadings of 10, 25, 50, and 70 wt.-% of the original silica gel mass (20 g) were tested and the results obtained are presented in Figure 6.18. When the adsorbent was loaded with 5 wt.-% MPO, it was assumed that this oil amount was insufficient to saturate the amount of adsorbent used, what was proved by the adsorbent color after the oil adsorption (light orange).

The highest loading (70 wt.-%) investigated in this work was beyond the maximum adsorption capacity, since the used silica gel was not able to adsorb completely that amount of oil. In the literature, the maximum adsorption capacity of citrus oils for silica gel was reported as 84 wt.-% [21]. As expected, the desorption at higher loadings proceeded faster than at lower loadings, what can be observed in Figure 6.18. This effect was due to the curved shape of adsorption isotherms, which resembled a Langmuir-type curve, as presented by Reverchon [123], Subra et al. [21] and Sato et al. [130]. At higher solute concentrations the relative concentration in the solute increased. Therefore, 25 and 50 % were considered as the optimum loadings for the deterpenation of red MPO with SC-CO_2.

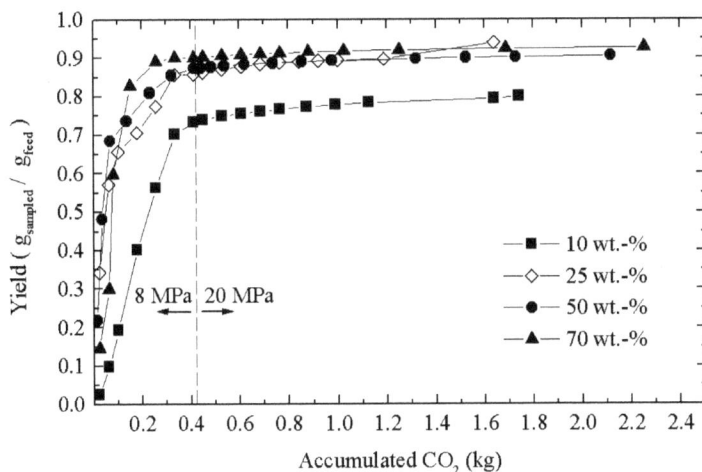

Figure 6.18. Effect of different loadings in the MPO desorption at 8 and 20 MPa and 40 °C.

Results obtained by Reverchon [123] at 10 wt.-% loading for a model system composed by limonene and linalool led to the assumption that the desorption selectivity decreases when the solute loading increases. In order to investigate the loading effect in the desorption between terpenes and aroma components, samples obtained for the experiments showed in Figure 6.18 were analysed. All identified terpenes and oxygenated aroma components were then separated in two different groups (terpenes and aroma). Non-identified components, especially low-boiling components like waxes, coumarins and flavonoids, were also present in the samples and were not considered in this analysis. Therefore, total terpene and aroma concentrations were normalized.

Figures 6.19 and 6.20 present the results obtained for the experiments carried out at 40 °C, 8/20 MPa and with a solvent flow rate of 0.41 kgCO_2/h. At higher loadings (50 and 70 %), larger extraction times were required to achieve total oil recovery (Figure 6.19). In order to provide a better visualization and evaluation of the loading effect, the most representative points of Figure 6.20 were chosen as the ones included in the interval 30-100 minutes. The pressure was increased up to 20 MPa after one hour. After this time, complete deterpenation was achieved for the experiments performed with 25 % loading. The other experiments achieved complete elimination of terpenes after approx. 80 minutes. Based on these experiments, 25 wt.-% was chosen as the best loading condition, since the complete elimination of terpenes could be achieved at shorter extraction times, what is directly related to lower solvent consumptions.

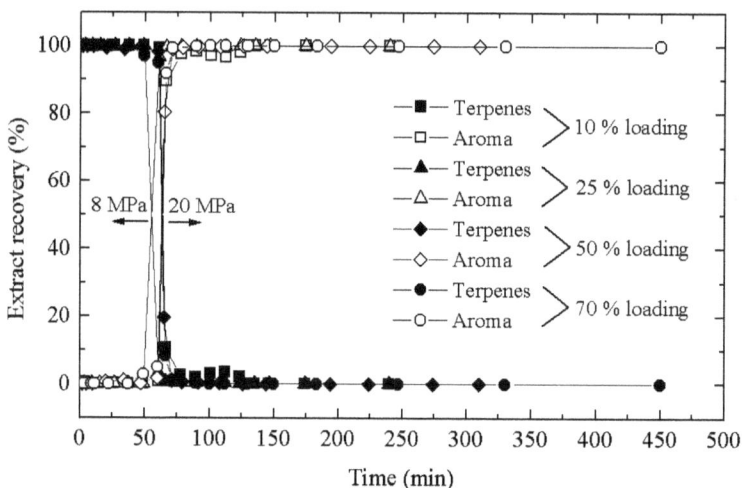

Figure 6.19. Effect of different MPO loadings in the desorption of terpene and aroma components at 8 and 20 MPa and 40 °C.

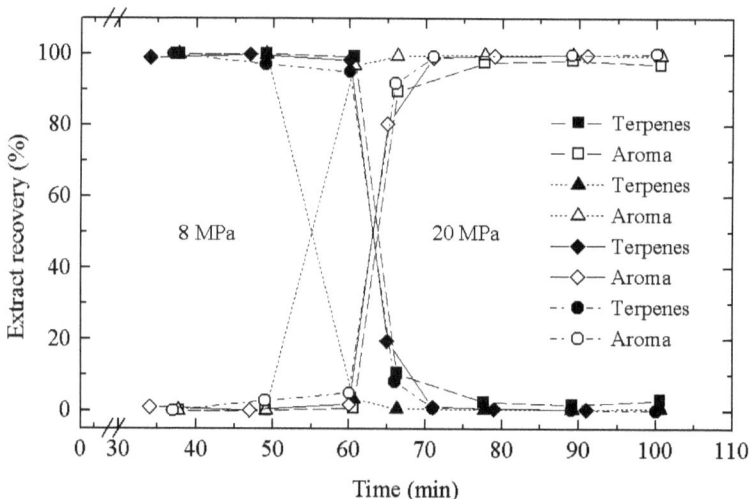

Figure 6.20. Effect of different loadings in terpene and aroma desorptions at 8 and 20 MPa and 40 °C (squares: 10 %; triangles: 25 %; diamonds: 50 %; circles: 70 %).

6.4.2.2.2. Effect of Pressure

Experiments at 35 and 40 °C were carried out with initial pressures of 7.5, 8, and 10 MPa and final pressures of 15, 20, and 24 MPa. As presented in Chapter 4.3.1, publications related to citrus oil deterpenation by ad-/desorption have chosen initial pressures in the range between 7.5 and 9.0 MPa (mostly at 7.5 MPa), while for the aroma desorption (second step) pressure ranges were higher, varying from 8.5 to 30 MPa.

Figure 6.21 shows the total terpene and aroma compositions for the performed experiments. Once more it must be mentioned that components summarized as "others" were mainly oxygenated ones, but some sesquiterpenes, alkanes and other low boiling unidentified components were also included in this group. The diagram shows that the fractions obtained at 40 °C and 8 MPa contained mostly terpenes, with much lower amounts of "others" desorbed in comparison with the other conditions investigated. The combination of a lower temperature of 35 °C with a higher starting pressure of 10 MPa (CO_2 density of 713.85 kg/m^3) was disadvantageous, since considerable amounts of decanal were desorbed along with the terpene components.

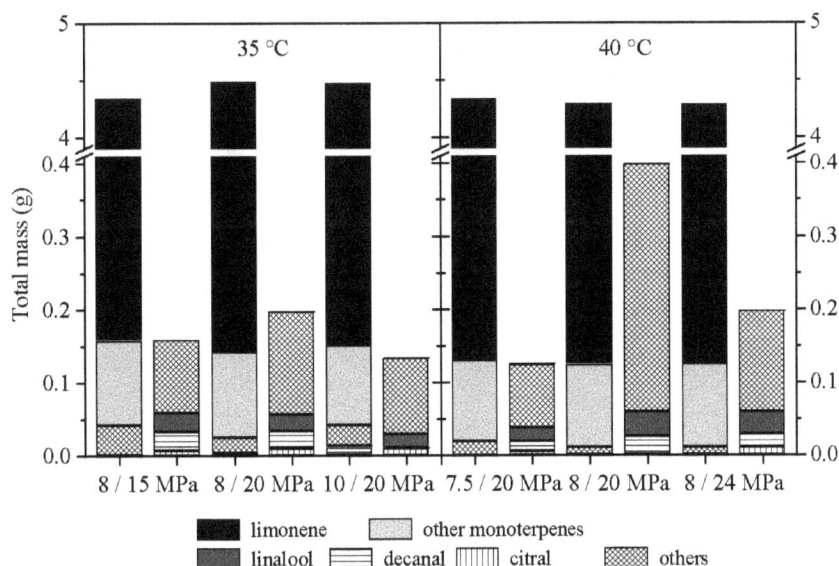

Figure 6.21. Total composition of terpene and aroma fractions obtained at 35 °C and 40 °C, 25 % loading and different pressure levels.

Additionally, higher amounts of linalool, decanal and "others" were obtained when using 8/20 MPa at 40 °C, proving that these conditions were more suitable for ad-/desorption of the red MPO. By increasing the pressure for the second step from 20 to 24 MPa at 40 °C, no change in the total amount of the most significant aroma substances could be observed. Using higher pressures without significant advantage should also be avoided because of the increased energy consumption, what would lead to higher operational costs. When using a pressure of 15 MPa instead of 20 MPa, the total recovery of aroma components decreased significantly.

The introduction of the Q factor has been suggested in order to quantify the relations between the masses of terpene and aroma components obtained. This factor was then defined as the ratio between the total amounts of limonene and linalool obtained at the first pressure level and the amounts of terpene and aroma components for the second pressure level. The values obtained at 40 °C are shown in Figure 6.22. Smaller Q factors are higly desired when analysing the second step of the desorption. Therefore, a combination of 8 and 20 MPa were then considered as the best desorption conditions for the system.

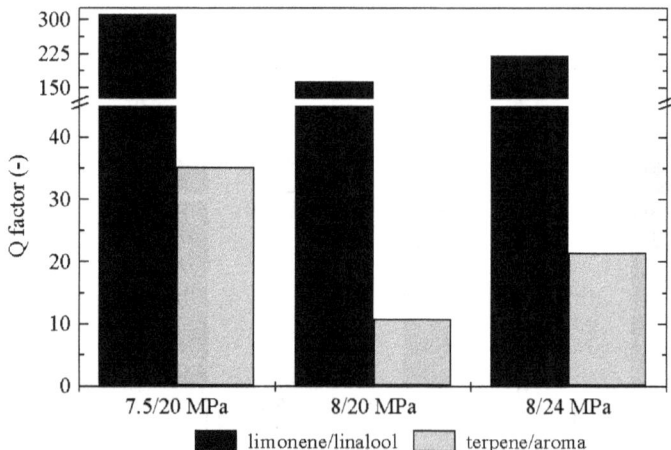

Figure 6.22. Q factors obtained at 40 °C for different pressures.

Last but not least, the recovery extraction curves obtained for the experiments mentioned above are presented in Figure 6.19. In these experiments, CO_2 density varied from 232.85 to 873.19 kg/m^3. As expected, when comparing the yield curves of the experiments performed at 35 °C and 40 °C, a much faster desorption could be observed for all experiments performed at

35 °C. However, due to the increase in the solvent density, desirable components (like decanal, see Figure 6.21, pressures of 10/20 MPa) were already desorbed in the beginning of the experiments. The selectivities obtained at 35 °C were lower than the ones obtained at 40 °C, providing a less effective separation at 35 °C, what will be discussed in the following Chapter.

6.4.2.2.3. Effect of Temperature

The temperature effect in the desorption of MPO has been also investigated. Experiments were performed at 35, 40, 50 and 60 °C and the extraction curves obtained are presented in Figure 6.23. When comparing all the overall extraction curves evaluated, it can be observed that the initial slopes at 8 MPa increased with decreasing temperature, caused by an increase in the CO_2 density, which varied from 192.18 kg/m^3 at 60 °C to 427.27 kg/m^3 at 35 °C. At 35 °C the solvent power of CO_2 was maximized, leading to the highest oil solubility obtained. This behaviour was observed for many substances at comparatively low pressures in several publications [21, 123, 164, 166].

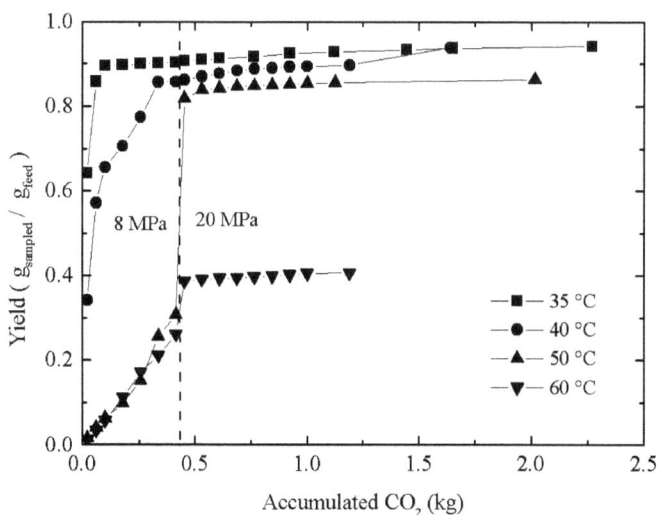

Figure 6.23. Evaluation of the temperature effect in the desorption of red MPO.

At higher pressures, when the density variation with temperature was lower, the increase in the temperature system provided an increase in the vapor pressure of the solute. Reverchon

[123] observed that, at a fixed pressure, an increase of the desorption temperature produced lower desorption rates of limonene due to the decrease in the CO_2 density. In this case, the solvent density effect was not counterbalanced by the increase in the vapor pressure of the solute. Temperatures above 40 °C were considered unsuitable because the pressure change to 20 MPa would lead to longer extraction times, what would cause higher CO_2 consumptions.

Figure 6.24 shows a comparison between the MPO composition profiles obtained at 35 and 40 °C. In these experiments, 5 g MPO were adsorbed in 20 g silica gel, corresponding to a loading of 25 %. The amount of oil desorbed after the consumption of 2 kg of solvent (0.41 kg/h) was practically the same in both cases, 4.7 g. The missed 0.3 g were assumed as negligible (oil still adsorbed in the solid or even normal process losses). At 8 MPa and 35 °C, the experiment was more effective in the removal of terpenes, what can be observed when comparing the masses obtained: 4.68 g against 4.29 g at 40 °C. However, the total amount of components desorbed at 40 °C after the pressure increase was twice higher. Concentrations of linalool, decanal and other aroma components were also higher at 40 °C, what can be probably explained by the counterbalance found between the solvent density and vapor pressure effects. Therefore, 40 °C was chosen as the best temperature condition investigated for the desorption of MPO.

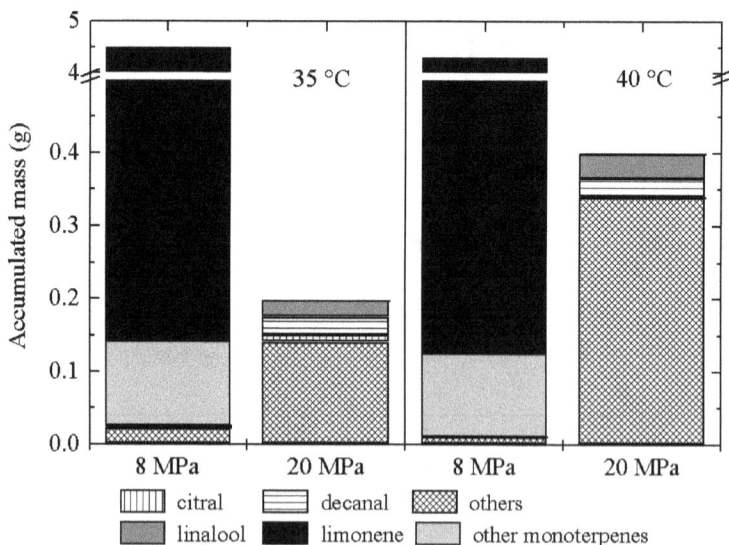

Figure 6.24. Evaluation of the Spanish MPO composition profiles obtained at 35 and 40 °C.

In order to completely observe the temperature effect on the oil desorption, the next step corresponded to the evaluation of the desorption course for the most important components at 40 °C, as presented in Figure 6.25. At 8 MPa, it can be observed that monoterpenes were practically totally desorbed. Very small amounts of some aroma components were desorbed at the same time. Due to their low mass percentage, they were only detectable after monoterpenes have been completely depleted, being considered then as negligible. At 20 MPa, decanal and most other aroma components were desorbed quite fast, while the amount of linalool in the samples decreased after reaching a maximum of 7.5 mg (9 wt.-% of the sample). This can be better explained by the higher slope of the linalool adsorption isotherm in comparison with other oxygenated compounds, what was also observed by Subra et al. [21]. Sinensal and its corresponding alkane 2,6,10-trimethyl-2,6,9,11-dodecatraene were desorbed very slowly, mainly due to their polarity and high molecular weights (see Table 4.3). As can be seen, sinensal was the only detectable aroma component in the last sample.

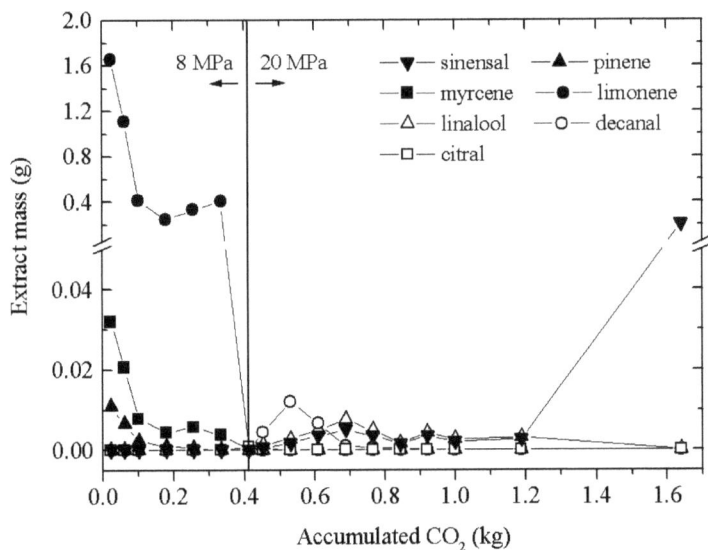

Figure 6.25. Course of the desorption of the most representative oil components at 40 °C.

6.4.2.3. Experiments with Green Brazilian Oil

Experiments with green Brazilian MPO were performed at the best determined process conditions for Spanish red oil, namely 40 °C and 8 and 20 MPa, with loadings of 25 % and 50 %. The results shown in Figure 6.26 indicate that the chosen conditions were also feasible for desorption of the Brazilian green oil, presenting also a comparison with the extraction recovery curves obtained for the Spanish red oil. The curves have presented the same trend previously reported for red MPO.

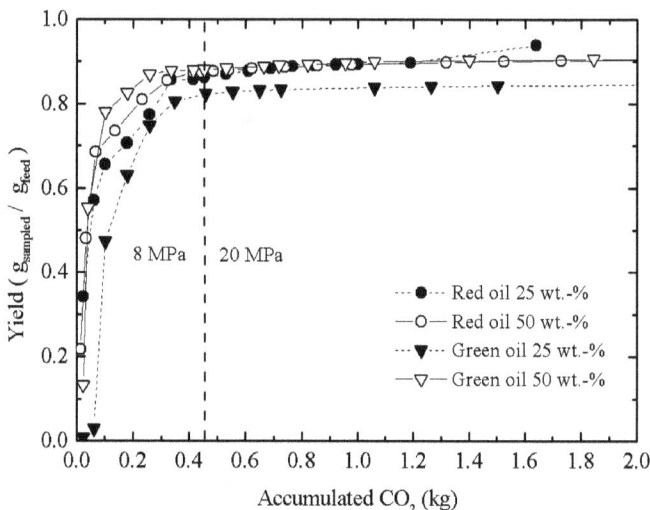

Figure 6.26. Extraction yield curves obtained for the red and green oils at 40 °C.

Figure 6.27 shows the composition profiles obtained for experiments with green oil carried out at 40 °C and with loadings of 25 and 50 %. Additionally, results obtained in the desorption of red oil are presented. When comparing the experiments in relation to the total amounts desorbed, it is clear that the desorption of the red oil achieved the best oil recovery. This can be easily explained because green oil components were not completely removed from the adsorbent. Even at larger extraction times, some low boiling green oil components were still adsorbed in the silica gel, what could be visually observed.

The higher percentage of other monoterpenes in the terpene fraction obtained for the green oil was due to the presence of terpinolene, which was not found in the red oil. The differences in the amounts obtained at 8 MPa were a function of the differences in the composition of the crude MPO used (see Table 6.1 for further comparisons). By analysing the Q factor for terpene/aromas as described in Chapter 6.4.2.2.2, it was possible to find the values of 22.7 and 34.0 for the experiments with 25 and 50 % loading, respectively. Once more the loading of 25 % provided the best results, achieving successfully the deterpenation of the oil. Thus, the separation performance could be considered comparable to that of the red Spanish oil.

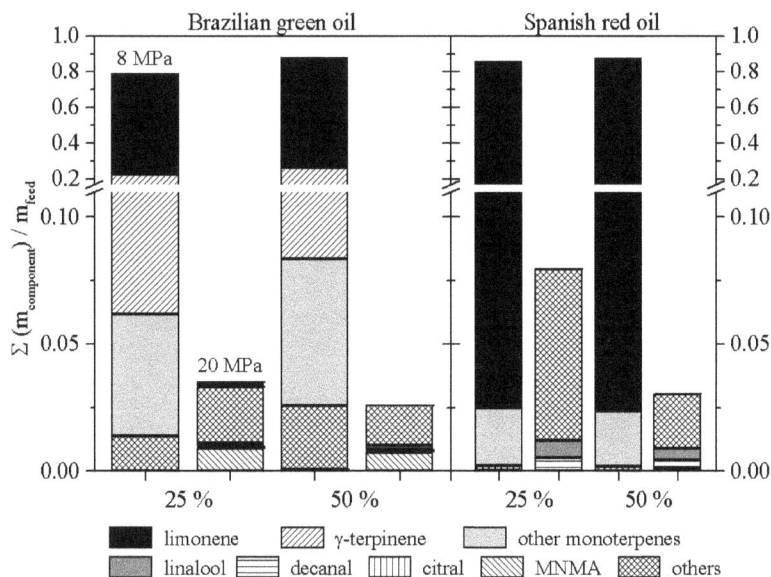

Figure 6.27. Total composition of terpene and aroma fractions obtained at 40 °C and 25 and 50 % loading for both MPO used in this work.

The extract recovery of one experiment carried out at 40 °C and 25 % loading is presented in Figure 6.28. The unidentified components are not presented. After 80 minutes, it can be observed that monoterpenes were practically totally desorbed. As previously observed during experiments with red oil, small quantities of some aroma components were simultaneously desorbed. After the pressure step, decanal, citral, linalool and especially MNMA were desorbed in relatively longer times. MNMA, decanal and linalool have reached maximum recoveries at different times, achieving approximately 5, 7 and 47 wt.-% of their respective samples. In the end of the extraction, citral and linalool were the only identified aroma components. Waxes, flavonoids and other unidentified low boiling components were assumed to be present in these samples.

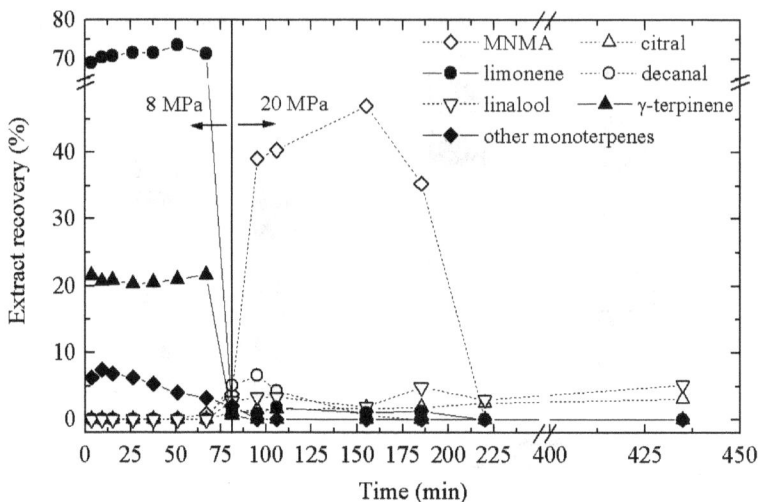

Figure 6.28. Desorption course of the most representative green oil components at 40 °C.

6.4.2.4. Desorption Residues – Regeneration of the Adsorbent

As mentioned in the previous Chapter, the adsorbent was still colored after the experiments with both MPO samples, especially with the green oil. The observed colors were lighter than before the desorption started, but still clearly visible. In order to determinate the remaining adsorbates and to perform the regeneration of the adsorbent, the silica gel was washed with warm ethanol as presented in the literature [127]. 250 ml of ethanol were heated up to 50 °C and used to wash the adsorbent. The resulting ethanol solution was concentrated in a rotary evaporator under vacuum. The solvent was evaporated and the solid material obtained was then analysed by GC and GC-MS in order to identify the restant components. These analysis were not able to identify possible oxygenated components still present in these samples: the residues contained several components, including carboxylic acids and their respective esters, terpenoids and other unidentified low boiling components, which ones were assumed to be waxes and flavonoids.

6.4.2.5. Process Scale-Up

The laboratory scale experiments conducted produced promising results. Thus, in order to scale-up the process, desorption experiments were carried out at 40 °C with both oils in a pilot plant comprising an extraction column of 1.5 liters. The pressure levels investigated were 8 and 20 MPa. Initially, the mass of silica gel to be used in the experiments should be evaluated.

Table 6.8 shows the results obtained for red MPO desorption with different adsorbent masses at 40 °C and 25 % loading at the lowest pressure level. The concentrations of the unidentified low volatile components (rests) were not considered. The identified components were grouped in terpene and aroma fractions, and their concentrations were then normalized. For these experiments, the best recoveries were obtained when using 80 g silica gel and a CO_2 flow rate of approximately 1.40 kg/h, corresponding to a solvent-to-feed ratio of 17.50. For 100 and 120 g of adsorbent, the solvent flow rates were approximately 1.75 and 2.10 kg/h. When using higher adsorbent masses, aroma components were desorbed simultaneously with terpenes at the lowest pressure, what was undesired. Therefore, a mass of adsorbent of 80 g and a loading of 25 wt.-% were chosen as the best conditions and were used in all the following scale-up experiments.

Table 6.8. Recovery of terpenes and aroma fractions with different adsorbent amounts.

Mass of adsorbent (g)	8MPa		20 MPa	
	Terpenes (wt.-%)	Aroma (wt.-%)	Terpenes (wt.-%)	Aroma (wt.-%)
80	100.0	0.0	0.0	100.0
100	99.0	1.0	0.9	99.1
120	98.0	2.0	1.5	98.5

Figure 6.29 shows the extraction curve obtained for one desorption experiment carried out for the conditions mentioned above with red Spanish MPO. The unidentified components are not presented. The results shown are quite similar to the ones obtained with the experiments carried out with the laboratory scale equipment. The oil deterpenation was achieved after 150 minutes, what corresponded then to the pressure increase point. This time corresponded to an approximately 2 times longer process in comparison with the laboratory scale experiments.

The relative amounts of aroma components recovered were in the same range, proving that the deterpenation of red MPO could be scaled up effectively.

At 20 MPa, aroma components have been desorbed slowly, since the recovery of decanal, linalool and sinensal increased 50 minutes after the pressure step, what corresponded to approximately 200 minutes. Concentrations of linalool and decanal increased up to 220 minutes, reaching maximum values of 2.5 and 1.5 wt.-%, respectively. After 300 minutes, these values have been changed to 0.5 and 0.2 wt.-%. As observed previously (Chapter 6.4.2.2.3), similar behavior for sinensal could be observed, which was the last component to be desorbed, reaching up to 13 wt.-% in the last collected sample.

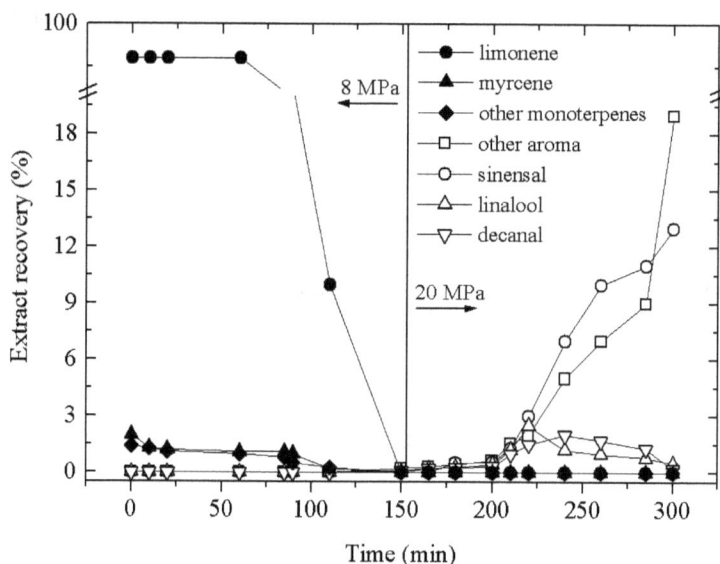

Figure 6.29. Scale-up results obtained for the desorption of red MPO (25 % loading).

For the green oil desorption, experiments have been performed and the results obtained are presented in Figure 6.30. At 40 °C and 25 % loading, the results were similar to the ones obtained previously (see Chapter 6.4.2.3). The unidentified components are not presented. It can be observed that, after 170 minutes, monoterpenes were totally desorbed, only traces were present. As presented in Figure 6.28, decanal, citral, linalool and especially MNMA were desorbed in relatively longer times after the pressure increase. Decanal, MNMA and linalool reached maximum recoveries between 270 and 300 minutes, achieving approximately 10, 38 and 5.25 wt.-%. In the end of the process, citral, linalool and MNMA were the only identified aroma components, with concentrations of 0.5, 0.2 and 5 wt-.%, respectively.

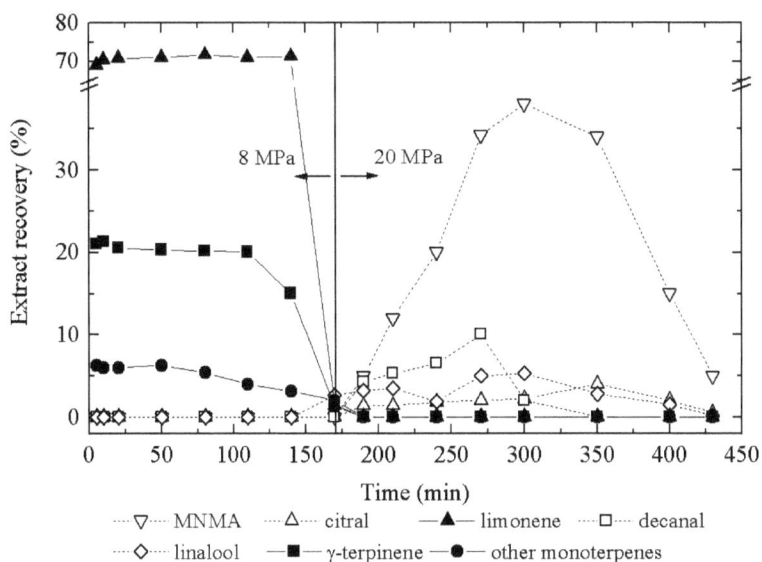

Figure 6.30. Scale-up results obtained for the desorption of the Brazilian MPO.

7. Fractionation and Refining of Rice Bran Oil with SC-CO$_2$

In this Chapter, the results obtained for the fractionation and refining of rice bran oil (RBO) with SC-CO$_2$ through solid batch extractions and countercurrent multistage fractionation are presented. Rice bran samples were supplied by Müllers Mühle (Gelsenkirchen) and Oryza (Hamburg). As soon as the rice bran samples were received, they were thermally stabilized by keeping them at 120 °C for approx. 24 hours, as discussed in Chapter 3.

The first sets of experiments consisted on RBO batch extractions. Experiments were carried out using pressures between 10 and 40 MPa and temperatures ranging from 40 to 60 °C. Additionally, the overall extraction curves (OEC) were modeled considering the crude RBO as a complex mixture. The fractionation of different classes of components, namely free fatty acids (FFA), triglycerides (TG), sterols and oryzanols was modeled. The used models were previously presented in Chapter 2.5.

Following the scope proposed for this work, results obtained through phase equilibria measurements are presented. Based on the vapor-liquid equilibria (VLE) data for the system, countercurrent extraction experiments were conducted and the multistage RBO fractionation could be evaluated.

7.1. RBO Extraction and Fractionation

Sequential RBO extraction experiments have been carried out in the laboratory scale equipment described in Chapter 5.2.1. Sets of experiments were conducted at pressures varying from 10 to 40 MPa and temperatures of 40, 50 and 60 °C.

In order to evaluate the best conditions for the extraction of crude RBO as a pseudo-component, a set of experiments was carried out at 50 °C and the results obtained are presented in Figure 7.1. These experiments were carried out at the maximum CO$_2$ flow rate provided by the equipment, which was 4.16 g/min. The highest extraction yield (22.3 wt.-%, g$_{oil}$/g$_{feed}$) was obtained at the highest operational pressure, 40 MPa. The extraction yield ranges achieved are comparable with the results presented in the literature. As expected, this is related to the higher solvent density at this condition (924 kg/m^3), what provided a faster penetration in the solid matrix and consequently a higher solubilization of the oil components in comparison with the other conditions. CO$_2$ density at 10 MPa was only 386 kg/m^3, what justified the lower extraction yield obtained at this condition during a six hours extraction.

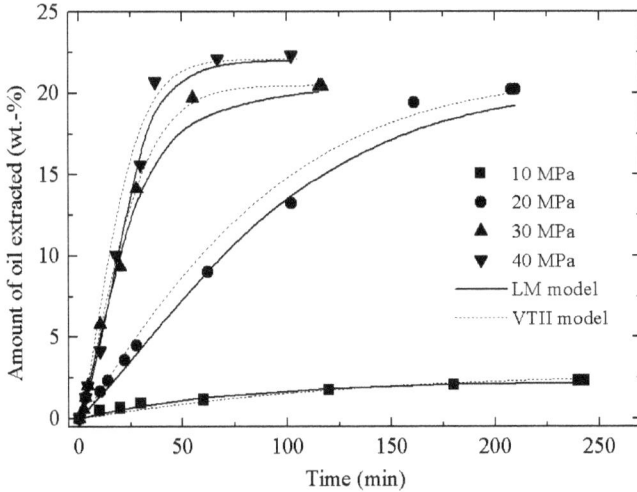

Figure 7.1. OEC obtained for crude RBO extraction at 50 °C and modeled curves.

Additionally, the experimental points in Figure 7.1 were modeled with good accuracy by employing the logistic and the VTII models, denoted as LM and VTII, respectively. The parameters used for the modeling with the VTII model are listed in Table 7.1. Dimensions of the extractor were 20 cm length and 13.6 mm inner diameter. The extractor was loaded with 19 g of rice bran and the average diameter of the particles was 0.72 mm.

Table 7.1. Data for the simulation using the VTII model.

Parameter	Unit	10 MPa	20 MPa	30 MPa	40 MPa
ε	-	0.65	0.65	0.65	0.65
k_1	-	2.0×10^{-2}	2.5×10^{-2}	1.0×10^{-1}	1.25×10^{-1}
D_{AX}	m^2/s	2.7×10^{-6}	1.39×10^{-6}	1.28×10^{-6}	1.21×10^{-6}
β	m^2/s	1.57×10^{-4}	1.23×10^{-4}	1.18×10^{-4}	1.15×10^{-4}
Re	-	10.69	4.39	3.54	3.08
Sc	-	6.99	8.35	9.35	10.14
Sh	-	10.78	8.44	8.11	7.91
Pe	-	0.23	0.22	0.22	0.22

The course of the RBO fractionation obtained at 20 MPa and 50 °C is presented in Figure 7.2. Initial concentrations of FFA (represented as a sum of myristic, palmitic, oleic and linoleic acids) were in the range of 8 wt.-%. Linoleic acid (C$_{18:2}$) was the most abundant one, followed by oleic (C$_{18:1}$), palmitic (C$_{16:0}$) and myristic (C$_{14:0}$) acids. FFA concentration profiles decreased slightly, achieving a value lower than 1 wt.-% after two hours. However, the same trend was not presented by the other components. Concentrations of sterols (campesterol and sitosterol) and oryzanols increased as soon as FFA concentrations started decreasing. Between 70 and 100 minutes, concentrations of campesterol, sitosterol and oryzanols achieved 0.28, 0.16 and 0.13 wt.-%, respectively.

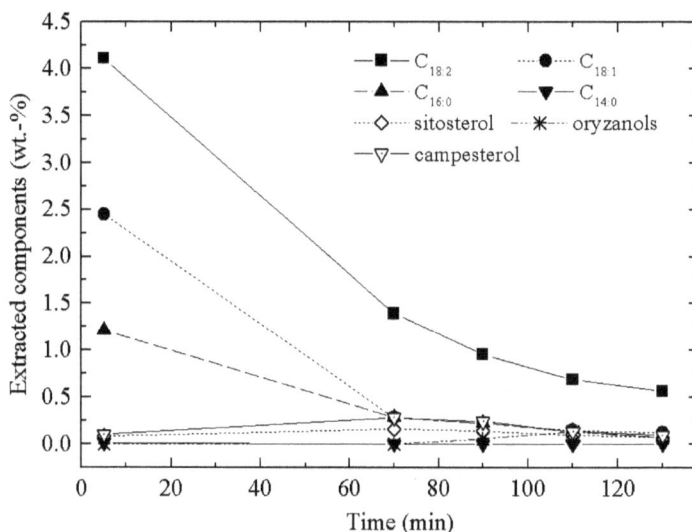

Figure 7.2. Evaluation of the amount of RBO key-components profile.

In order to better evaluate the course of the fractionation, data obtained at 20 MPa and 50 °C were presented in function of relative concentrations. Figure 7.3 shows the relative concentrations (C/C$_0$) of each key-component evaluated. Relative concentrations of linoleic acid practically remained constant during the experiment, whereas the relative concentrations of the others FFA decreased. On the other hand, concentrations of sterols and oryzanols increased slightly, up to 10 times the initial concentrations. This behavior was also observed in previous works [52, 59, 91, 93, 94], with variations varying from 5 to 8 times the initial concentrations.

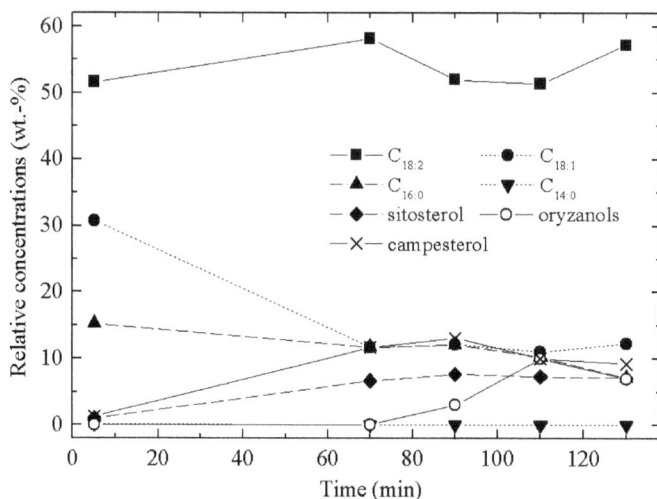

Figure 7.3. Relative concentrations obtained at 20 MPa and 50 °C.

As suggested by previous works [91, 93], the separation between FFA, sterols, oryzanols and TG can be more effectively achieved when employing lower solvent densities, i. e., when higher temperatures are employed. In order to investigate the effects of temperature and pressure on the selectivity between the key-components evaluated, the parameter M was introduced. M was defined as the ratio between the total mass of component i and the total mass of component j obtained under the same conditions, including extraction time, which one was estipulated as 100 minutes. The obtained results are presented in Figure 7.4. In this analysis, component i was fixed as FFA, since FFA were the undesired components to be firstly fractionated. The other components were then defined as component j: TG, sterols (ST) and oryzanols (OR). The highest M values were obtained between FFA and sterols (FFA/ST) under all investigated conditions. This was expected, since the amounts of sterol analysed after 100 minutes extraction were much lower than the others: sterol masses varied between 15 and 24.4 mg. For FFA, TG and oryzanols, the total mass amount after the same time period varied from 363 to 520 mg, 2 to 3.75 g, and from 31 to 68.8 mg, respectively. It can be clearly observed that the highest values for the M factor were obtained at 60 °C for all pressures evaluated, what can be observed by comparing the diagrams presented. At 25 MPa and 60 °C, maximum values for $M_{FFA/TG}$, $M_{FFA/ST}$ and $M_{FFA/OR}$ were 0.235, 31.33 and 15.16. In addition, a trend was observed for experiments carried out under the same temperatures: the lower the pressure, the higher were the obtained M factor values.

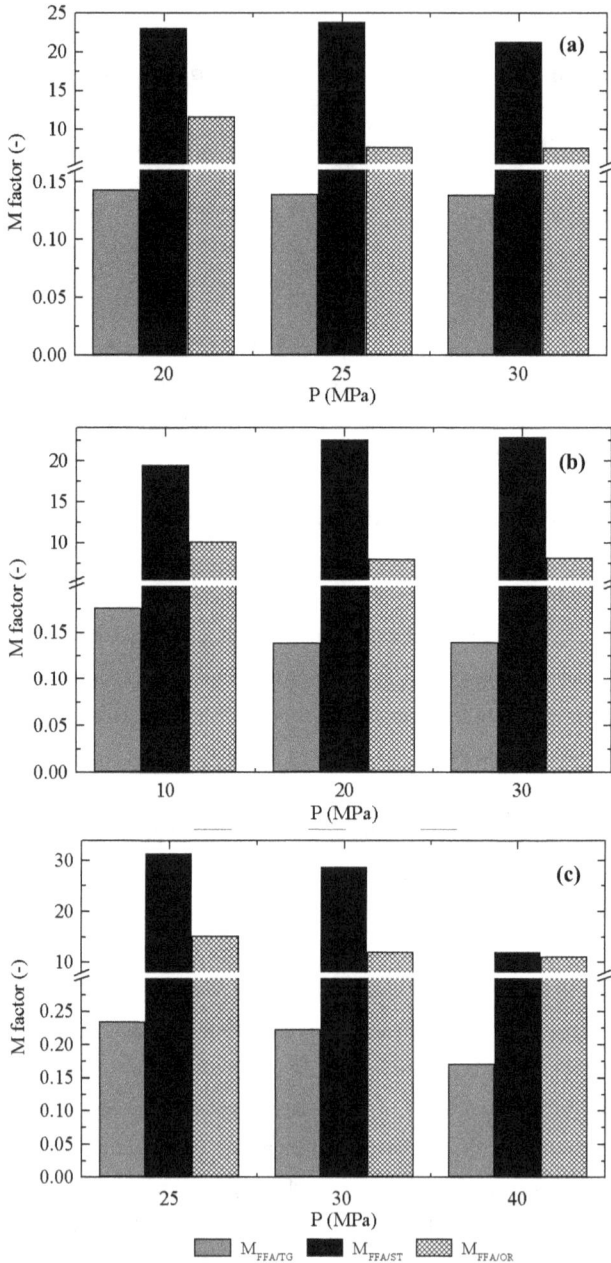

Figure 7.4. M factors obtained at different conditions: (a) 40 °C; (b) 50 °C; (c) 60 °C.

7.1.1. Modeling the RBO Fractionation

In order to evaluate the applicability of two different models presented in the literature [1, 22] for the fractionation of RBO, individual components were grouped in four different classes, represented by FFA, TG, sterols and oryzanols. The groups evaluated accounted for approximatelly 97 wt.-% of the total crude RBO samples used.

Based on the previous fractionation evaluation (Figure 7.4), higher selectivities between FFA and the other RBO components were obtained at the highest temperature investigated, which was 60 °C. Therefore, modeling was firstly performed for the experiments conducted at this temperature. Figure 7.5 shows the experimental results obtained at 25 MPa, including also the modeled curves using the LM and VTII models. Relative concentrations can be additionally denoted as C/C$_0$, corresponding to the degree of extraction obtained for each individual component within each analysed sample. A good accuracy was achieved when using both models, what can be observed by analysing Figure 7.5.

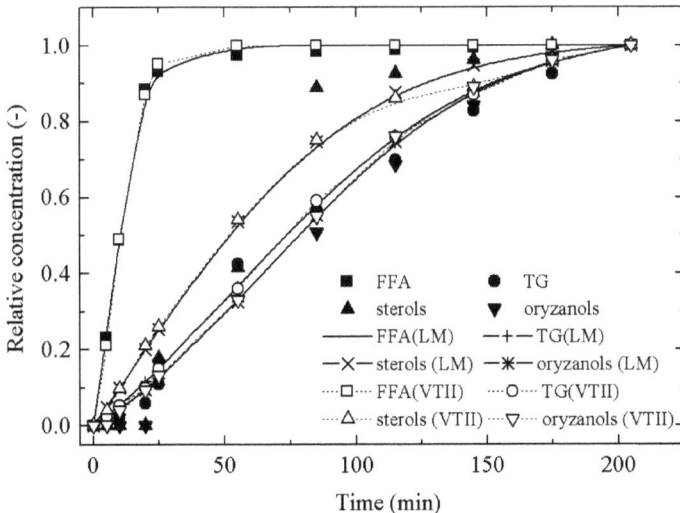

Figure 7.5. OEC obtained during the fractionation of RBO at 25 MPa and 60 °C.

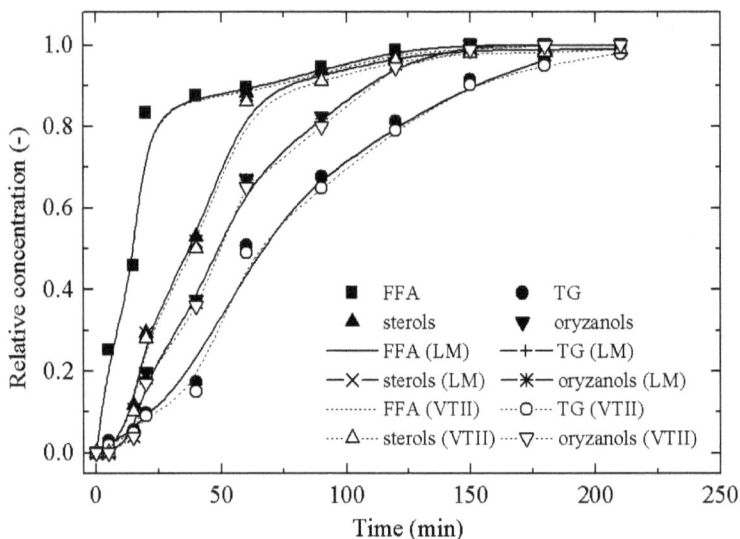

Figure 7.6. OEC obtained during the fractionation of RBO at 30 MPa and 60 °C.

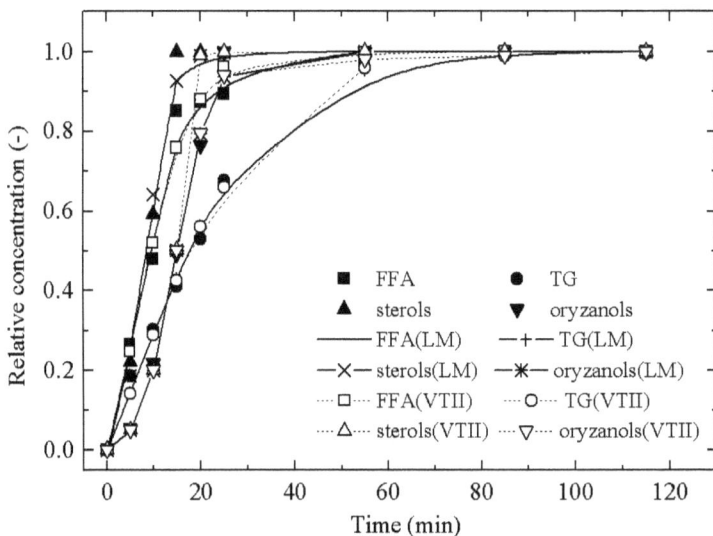

Figure 7.7. OEC obtained during the fractionation of RBO at 40 MPa and 60 °C.

Figures 7.6 and 7.7 show the OEC obtained and modeled at 60 °C and 30 and 40 MPa, respectively. When comparing Figures 7.5, 7.6 and 7.7, it is clear to observe that the higher the pressure, the less selective is the fractionation. At higher CO$_2$ densities (891 and 830 kg/m^3 at 40 and 30 MPa, respectively), the evaluated components were solubilized simultaneously, affecting consequently the separation efficiency in each sample.

At 25 MPa, the solvent density was 787 kg/m^3, proving to be a suitable condition for the fractionation of the undesired FFA from the other valuable nutraceutical components. The difference between the relative concentrations was confirmed by the larger distances between the FFA data and the other components. However, OEC for the low boiling components (TG, sterols and oryzanols) were very close to each other, suggesting that another extraction method, like countercurrent extraction or ad-desorption, could be employed in order to separate these components from each other.

A set of experiments at 50 °C was additionally performed. Pressures investigated were 25 and 30 MPa and the results obtained are presented in Figures 7.8 and 7.9. Analougsly to the observed from Figures 7.5, 7.6 and 7.7, a better fractionation between FFA and low boiling components was achieved when employing 25 MPa than at 30 MPa. However, the behavior of the OEC of sterols and oryzanol was different in comparison with OECs obtained at 60 °C. This can be probably explained by the influence of the vapor pressure in the solubilization of these components.

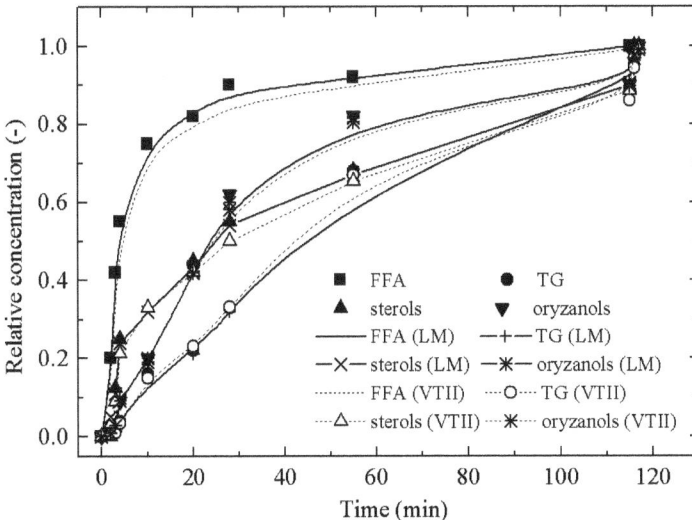

Figure 7.8. OEC obtained during the fractionation of RBO at 25 MPa and 50 °C.

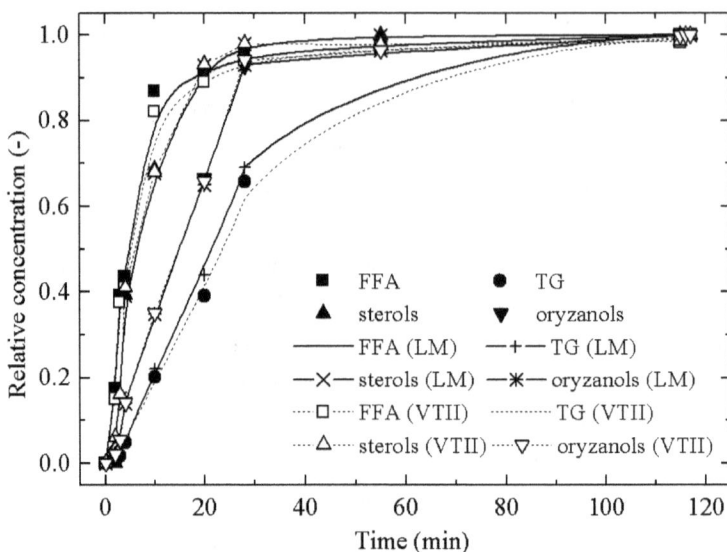

Figure 7.9. OEC obtained during the fractionation of RBO at 30 MPa and 50 °C.

The temperature effect on the RBO fractionation can be easily observed when comparing Figures 7.5 and 7.6 with Figures 7.8 and 7.9, respectively. It is clear that, at 25 MPa and 60 °C, fractionation between FFA and the other components can be more efficiently achieved. The same could be concluded when comparing the experiments carried out at 30 MPa, observing once more the higher selectivities reached at lower solvent densities.

7.2. RBO Refining

As already discussed in Chapter 3.2, the definition of refining is very ambiguous, providing different denotations depending on the country the processes are carried out. Normally, it refers not only to the removal of nontriglyceride fatty components from an edible oil, but also to neutralization, bleaching and deodorization steps. When dealing with the removal of FFA from RBO, the most used terms found in the literature are deacidification and refining, since this process step corresponds to the more delicate one in the oil production process. Previous works on the deacidification of RBO did not employ external reflux in order to enrich the most volatile components in the extracts; they have used the countercurrent columns as stripping ones, whereas others works applied temperature gradients within the column, inducing then the formation of an internal reflux [22, 54, 70, 99-101, 167].

In the following Chapters, the results obtained for the refining/deacidification of RBO with supercritical carbon dioxide are presented. Initially, the characterization of the oil samples used is presented.

7.2.1 . Characterization of RBO Samples

RBO samples used for the refining/deacidification experiments were extracted using the largest extraction apparatus available (extractor capacity of 15.8 liters), as described in Chapter 5.1. For these experiments, rice bran particles were previously agglomerated using a drum agglomerator. Extraction conditions were 30 MPa and 60 °C, achieving extraction yields between 19 and 22 wt.-%. As soon as the samples were obtained, they were kept under refrigeration and inert atmosphere (nitrogen) until they were used in the experiments.

The average crude RBO concentration profile obtained through the analytical methods used (see Chapter 5.1.1) is presented in Table 7.2 and was used as the average feed material composition for all following RBO experiments. When comparing the compositions obtained with the ones available in the literature, concentration for oryzanols, sterols and TG were in the same ranges presented by several authors [54, 94, 96]. The FFA concentration was a little higher than expected, since several works mentioned concentrations around 5 and 7 wt.% [54]. It is well known that the FFA concentrations are dependent on rice cultivar, but in this case it can be also related to the rice bran stabilization technique employed. However, the value of 10 wt.-% was considered acceptable, since FFA concentrations up to 16.5 wt.-% have been presented in the literature [168].

Table 7.2. Composition of the crude RBO used as feed material.

Component	Mass (wt.-%)
FFA	10.00
Oryzanols	1.30
Sterols	0.48
TG	84.66
Non identified	3.56

7.2.2. Phase Equilibria Measurements

In order to evaluate the mutual solubility of the pseudo-binary system crude RBO/SC-CO$_2$ and the distribution of the major components between liquid and gas phases, phase equilibria measurements were performed at three different isotherms (40, 60 and 80 °C) within a pressure range from 10 to 30 MPa. Investigations on the system solubility provide important information for the separation analysis, which are related directly to the solvent-to-feed ratio to be employed, as well as to the effective height of the column.

The presentation of the phase equilibria behavior is done by using a P-x diagram. Figure 7.10 shows the dependence of the binary solubility for crude RBO and CO$_2$ at different pressure and temperature levels. As expected, the mutual solubility increases with increasing pressure. At the lowest temperature investigated (40 °C), solubility of the low volatile RBO increased from 0.03 wt.-% at 10 MPa up to 1.8 wt.-% at 30 MPa. It can also be observed that the increase in the temperature at constant pressure caused a decrease in the solubility of RBO in CO$_2$. This can be explained by the lower solvent density obtained at higher temperatures: at 25 MPa, the density of the solvent decreased from 880 (40 °C) to 687 kg/m^3 (80 °C), causing a decrease in the solubility from 1.2 to 0.58 wt.-%.

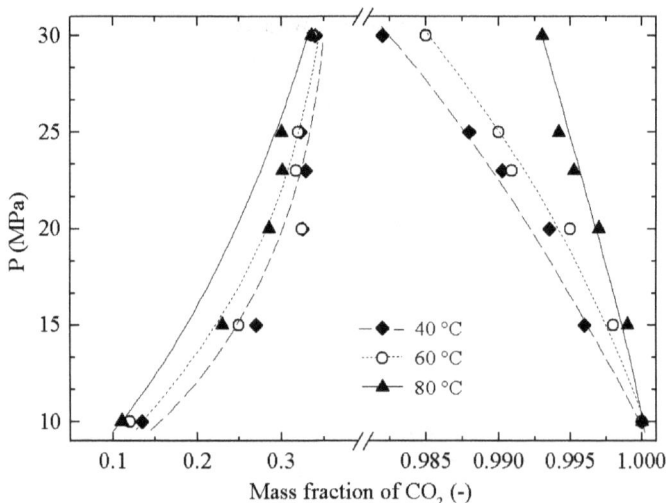

Figure 7.10. Mutual solubility of crude RBO and SC-CO$_2$.

The temperature increase is usually able to enhance the vapor pressure of the oil and consequently its solubility, but sometimes that can not be sufficient to overcome the effect of decreasing CO$_2$ density. Depending on the system pressure, the density or the vapor pressure effect can be dominant. When presenting the solubility results obtained by means of a ρ-x diagram, temperature and density effects can be better observed. Normally, only the vapor phase is used. Figure 7.11 shows the results obtained, when plotting the solute solubility in relation to the solvent density.

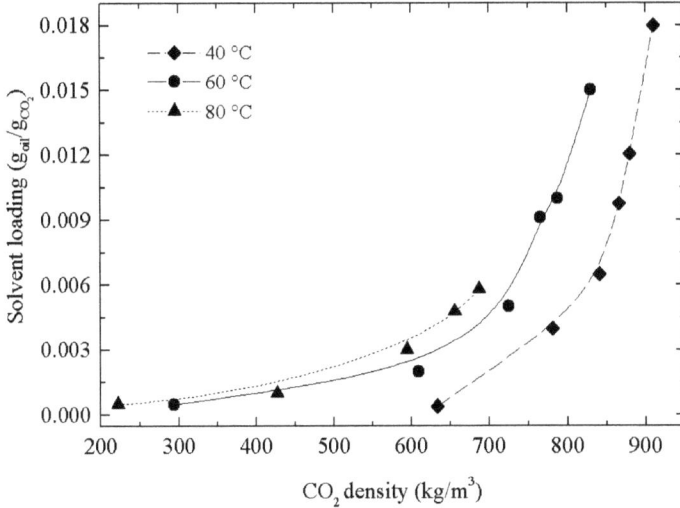

Figure 7.11. Solubility of crude RBO in dependence with CO$_2$ density.

From Figure 7.11 it is clear to observe that at determined solvent densities, the solubility of RBO increased with the temperature. At a density of 781 kg/m^3 at 40 °C, the CO$_2$ loading was around 0.4 wt.-%, while at 60 °C the solvent capacity increased to approximatelly 1.0 wt.-%, corresponding to 2.5 times the value obtained at the lower temperature. This comparison exemplifies the dominant effect of the solute vapor pressure over the decrease in CO$_2$ density when carrying out experiments at higher temperatures.

Table 7.3 presents the phase equilibria experimental values obtained. The measurements were calculated as an average value of four samples. Additionally, the reproducibility of the readings was confirmed, since standard deviation values obtained for gas phase measurements were slightly higher than the ones obtained for the liquid phase (approx. 0.017 against 0.008, in average) due to the lower amount of oil collected during sampling.

Table 7.3. Solubilites obtained for RBO in SC-CO$_2$.

T (°C)	P (MPa)	Density (kg/m^3)	Solubility liquid x_{oil} (g/g)	Solubility gas y_{oil} (g/g)
40	10	633	0.866	0.000
	15	781	0.730	0.004
	20	841	0.675	0.007
	23	866	0.672	0.010
	25	880	0.678	0.012
	30	910	0.660	0.018
60	10	293	0.880	0.001
	15	609	0.750	0.002
	20	725	0.676	0.005
	23	765	0.683	0.009
	25	787	0.680	0.010
	30	830	0.663	0.015
80	10	222	0.890	0.001
	15	427	0.770	0.001
	20	594	0.714	0.003
	23	656	0.699	0.005
	25	687	0.700	0.006
	30	746	0.665	0.007

After analysing the system crude RBO/CO$_2$ as a pseudo-binary system, the next step corresponded to the multicomponent treatment of the RBO samples. A pseudo-binary treatment is not adequate for the complete description of a multicomponent mixture like RBO. Some components present the capacity to be better dissolved in a gaseous phase than others, which remain in the liquid phase. In order to evaluate which components were enriched in the gas phase and which ones remained dissolved in the liquid phase, the distribution coefficients of the main components present in RBO (see Table 7.2) were investigated. The distribution coefficient can be calculated as the ratio between the mass fraction of one component in the gas phase (y_i) and its mass fraction in the liquid phase (x_i), as presented by Equation 7.1:

$$K = \frac{y_i}{x_i} \tag{7.1}$$

Figure 7.12 illustrates the results obtained through the calculations of the distribution coefficients K$_i$ for the main components of RBO, namely FFA, TG, sterols and oryzanols at 60 °C. Through the analysis of this diagram, it becomes very clear the tendency shown by the different classes of components: FFA were enriched in the gas phase (K>1), whereas TG, sterols and oryzanols remained dissolved in the oil phase (K<1). FFA were the most volatile components found, with K values varying from 4.2 to 3.18. In a countercurrent separation process, FFA would be extracted at the top of the separation column.

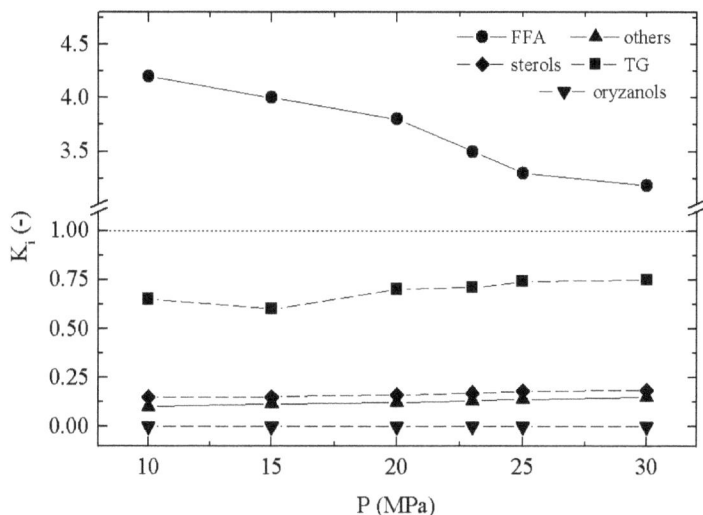

Figure 7.12. Distribution coefficients for main components of RBO obtained at 60 °C.

The other components evaluated presented distribution coefficients lower than 1, proving that, as expected, they remained dissolved in the oil phase. K values for TG were higher than the other components enriched in the liquid phase, varying from 0.65 to 0.75. Other components and sterols presented K values in the same range, with very small differences. Distribution coefficients obtained for oryzanols were very small ($\rightarrow$0), since not even traces of these components were detected in the gas samples.

On the basis of the data provided by Figure 7.12, it can be additionally concluded that, at lower CO$_2$ density (293 kg/m^3 at 10 MPa and 60 °C), the FFA separation from the other components (especially from TG) can be more efficiently performed. On the other hand, since TG, sterols and oryzanols presented close K values, the separation between them becomes more difficult, requiring further separation process steps. With higher K values differences, higher selectivities between the oil components are achieved. This was observed by Bamberger et al. [169], Maheshwari et al. [92], Jungfer [170] and Shen et al. [91, 93] in their previous works.

7.2.3. Separation Analysis

The separation analysis can be performed by employing shortcut methods. In this work the method of Jänecke was used. Considerations on this method have been presented previously in Chapter 2.3.

Based on the results obtained through the phase equilibrium measurements, the separation analysis could be performed. The high volatile components (FFA) present in the feed material (crude RBO) are better soluble in supercritical solvents that TG, sterols and oryzanols. Since the concentrations of sterols and oryzanols in the feed material (Table 7.2) were very low (around 1.8 wt.-%), the separation analysis was performed taking into consideration a pseudo-component composed by 89.44 wt.-% TG and 10.56 wt.-% FFA. These concentrations were obtained when normalizing the FFA and TG concentrations presented in Table 7.2.

Figure 7.13 shows the Jänecke diagram calculated at 25 MPa and 60 °C. The lines obtained for the extract and raffinate are presented. Additionally, the separation factors were obtained through the analysis of the distribution coefficients presented in Chapter 7.2.2.

The result of the separation analysis is presented in Figure 7.14. The Figure shows the dependencies of the theoretical number of separation stages (n_{th}, dotted lines) and the solvent to feed ratio (S/F, solid lines) on the reflux ratio (ν) obtained at 25 MPa and 60 °C. For this case, the separation efficiency was calculated for concentrations of 0.5 wt.-% FFA in the raffinate and 99.5 wt.-% FFA in the extract.

An increase in reflux ratio presents two confliting effects. Firstly, it results in a decrease in the number of stages necessary for the separation, which is highly desired and advantageous. This effect is very distinctive near the minimum reflux ratio, ν_{min}. With increasing ν, the necessary number of separation stages converges towards a minimum number of separation stages $n_{th,min}$. Consequently, a higher reflux ratio can result only in a small decrease of the number of theoretical stages.

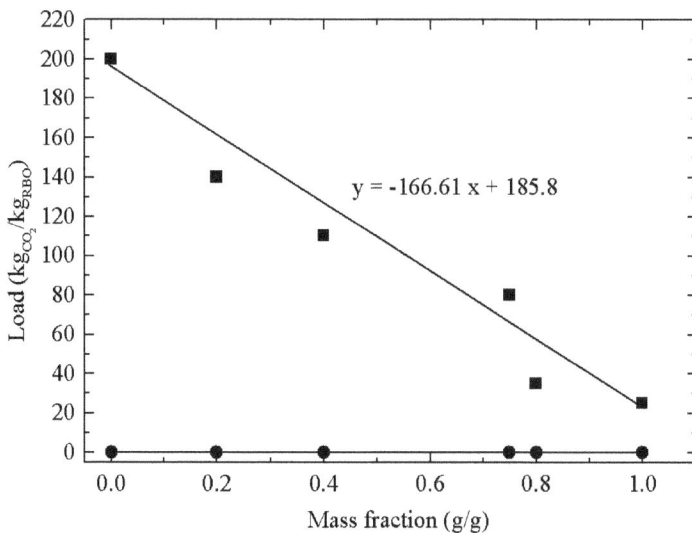

Figure 7.13. Jänecke diagram for RBO at 60 °C and 25 MPa.

On the other hand, a higher reflux ratio causes the S/F ratio to increases linearly. This must be avoided, since higher S/F values correspond to higher supercritical solvent consumptions, increasing operational costs. Thus, a compromise has to be found in respect to a number of stages being as small as possible, because this value directly influences the height of a column (investment costs), and a small S/F ratio (operational costs). For this case, the reflux ratio achieved its lowest value by 1.1, which corresponded to a minimum solvent-to-feed ratio of approximatelly 4.8 and the minimum number of theoretical stages was calculated to be 4.3. According to these calculations, approximatelly four equilibrium stages were achieved. Generally, performances of high pressure column packings allows a height equivalent to one theoretical stage (HETS) of 0.5–2m. Due to this, an industrial CC-SFE unit would need a total height of about 8 m for this specific separation task.

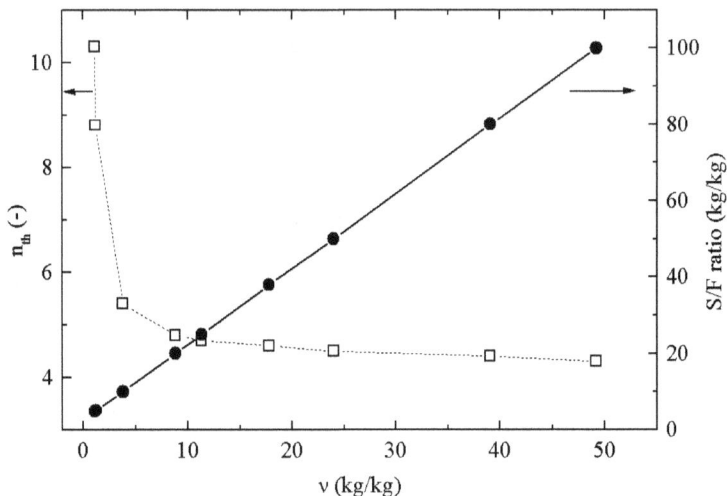

Figure 7.14. Calculation of the number of theoretical stages (25 MPa, 60 °C).

7.2.4. Countercurrent Experiments

Countercurrent experiments with crude RBO were carried out in two different ways. Initially, the experiments were performed without extract reflux. The apparatus used for this set of experiments was the one composed by a 6 m a height column (4 m EX packings) and 25 mm internal diameter, as presented in Chapter 5.2.3. After performing these sets of experiments, the next step corresponded to experiments using extract reflux, intending to achieve higher FFA enrichments in the extract fraction and a consequent enrichment of low boiling components (TG, sterols and oryzanols) in the raffinate. The second set of experiments was performed using the highest column available (7 m height and 17.5 mm ID), employing additionally extract reflux. The results obtained are presented in the following Chapters.

7.2.4.1. Experiments without Extract Reflux

Since most previous works have investigated the supercritical countercurrent fractionation of RBO without external extract reflux (see Chapter 3.3.2 for further details), the experiments presented in this Chapter were performed at one fixed temperature (60 °C) and at four different operational pressures (14, 17, 20, and 25 MPa) in order to compare the obtained results with the ones previously published by several authors.

When observing Figure 7.15, it can be seen that the solute loading of SC-CO$_2$ (g of extract collected/kg of CO$_2$) increased with increasing pressure under isothermal conditions. Data obtained at 60 °C were in good agreement with the data published by Dunford et al. [70]. In order to observe the tendency at different temperatures, data points plotted at 45 and 80 °C were taken from the same study. Additionally, a decrease in the temperature led to an increase in the solvent loading. This can be explained by the higher SC-CO$_2$ density at higher pressures and lower temperatures, achieving a higher solvent power under these conditions and leading to lower solvent requirements.

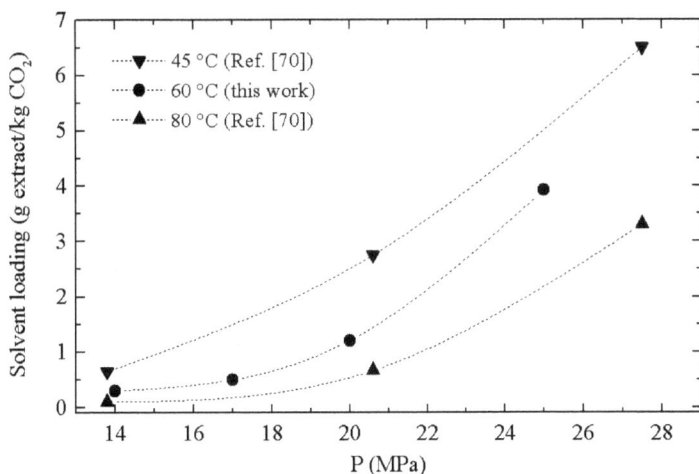

Figure 7.15. Temperature and pressure effects on the solvent loading.

From the literature [70], it is known that FFA removal from crude RBO is low at high pressures and low temperatures. That indicates that countercurrent fractionation under these conditions is not suitable for efficient FFA removal from the crude RBO. At higher CO$_2$ densities, the enhancement in the volatility of the low boiling components present in the mixture is more accentuated, achieving consequently higher TG, oryzanol and sterol losses in the extract fractions. Based on these findings, only one temperature level was investigated when performing experiments without extract reflux.

Figure 7.16 presents the effect of pressure in the FFA and TG compositions in extract and raffinate samples obtained at different pressure levels. The experiments were carried out for approximately 4 hours after steady state was reached. The FFA concentrations obtained were in good agreement with the values published by Dunford et al. [70]. Due to the little masses of extract samples collected during these experiments, the presence of some low boiling component in the extracts could not be confirmed even when applying higher pressures (20 and 25 MPa). As expected, masses of raffinate samples collected were much higher than the respective extract ones. No significant difference in the TG concentration in the raffinate could be observed.

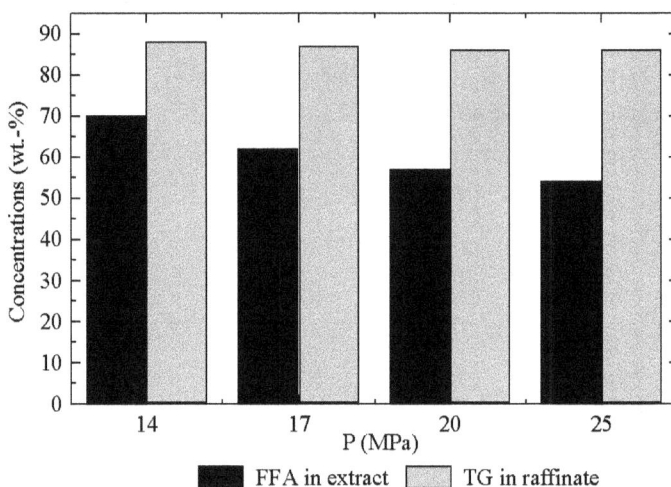

Figure 7.16. FFA and TG concentrations obtained at different pressures at 60 °C.

Figure 7.17 shows the effect of the cross section capacity on the composition of the extract at 60 °C and 14 MPa. The dotted lines represent the average FFA concentrations for this operational condition. Small variations in the FFA concentrations were observed (approximatelly ± 2 wt.-%) and were considered negligble, since the samples collected were in the range of some grams per hour and the analytical method used for their quantification could provide integration deviations up to 1 wt.-%.

Following the same trend, influences of the folding ratio on the extract composition were also considered negligble. This can be observed in Figure 7.18, where the FFA concentrations varied around ± 2 wt.-% at 60 °C and 14 MPa. When carrying out experiments under these conditions, smaller column heights would be needed, what was also observed by Budich [112] in the fractionation of orange peel oil with SC-CO$_2$.

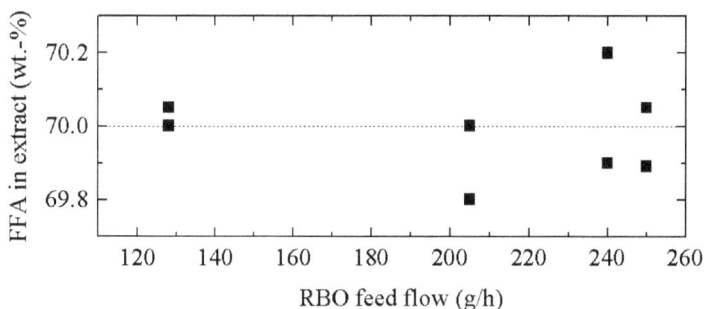

Figure 7.17. Effect of crude RBO flow on the FFA concentrations in extract samples.

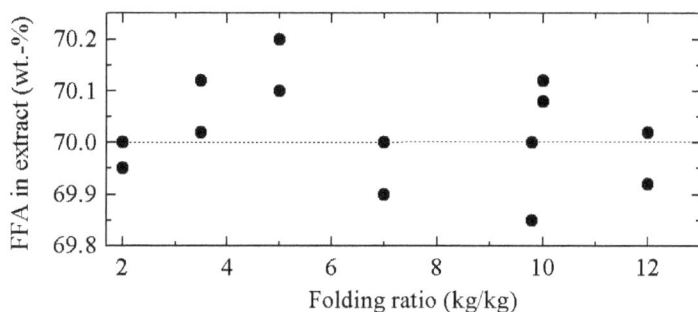

Figure 7.18. Effect of the folding ratio on the FFA concentrations in extract samples.

7.2.4.2. Experiments with Extract Reflux

In order to verify the influence of adding external extract reflux on the solvent loading (g extract/kg CO$_2$), experiments at 60 °C were performed at different reflux ratios. The results obtained are presented in Figure 7.19. At the highest reflux ratio, the highest solvent loadings were achieved, following the trend observed by other authors [112, 145]. At the highest pressure level investigated, it could be observed that the solvent loading slightly improved from 5 to 9, whereas its enhancement was proportionally higher between reflux ratios of 2 and 5. At 20 MPa, the highest improvement was achieved between reflux ratios of 5 and 9.

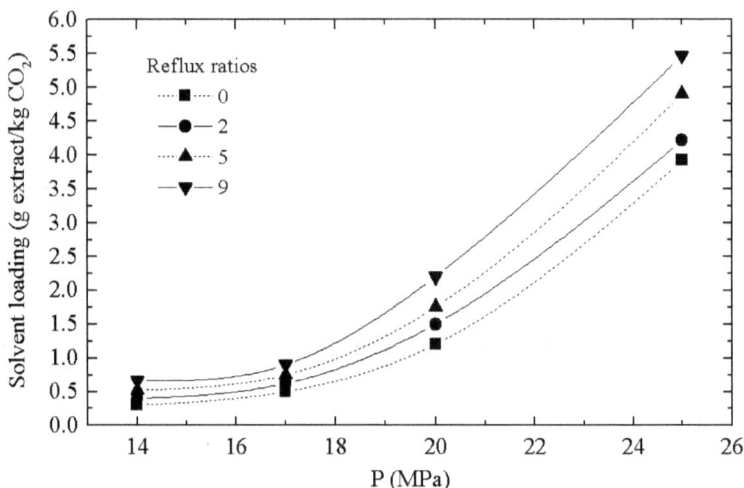

Figure 7.19. Influence of reflux ratio on solvent loading at 60 °C.

The influence of extract reflux on the extract composition was carried out at 60 °C at different pressure levels. Figure 7.20 shows the results obtained at constant extract reflux ratios. The dotted lines followed the trends obtained during the fractionation of mandarin peel oil with SC-CO$_2$ (see Figure 6.12) for the extract reflux ratios investigated. At higher reflux ratios, the trends obtained at the lowest solvent densities (14 and 17 MPa) showed the expected behavior. The highest improvements in the FFA concentrations were achieved when increasing the reflux ratio from 2 to 5, while when icreasing it from 5 to 9 only a slightly increase was observed. A reflux ratio of 9 was very difficult to be achieved due to the limitations of the extract pump and difficulties in the solvent regeneration.

The quality of the products obtained by countercurrent fractionation processes is directly proportional to the number of stages of the column, consisting on the main limiting factor for a determined separation task. In order to better evaluate the best operational conditions to achieve higher efficiencies in the deacidification of crude RBO, the solvent-to-feed ratio (denoted as SFR) becomes a more interesting operating parameter, since it can be directly related to the operational costs of a multistage fractionation process.

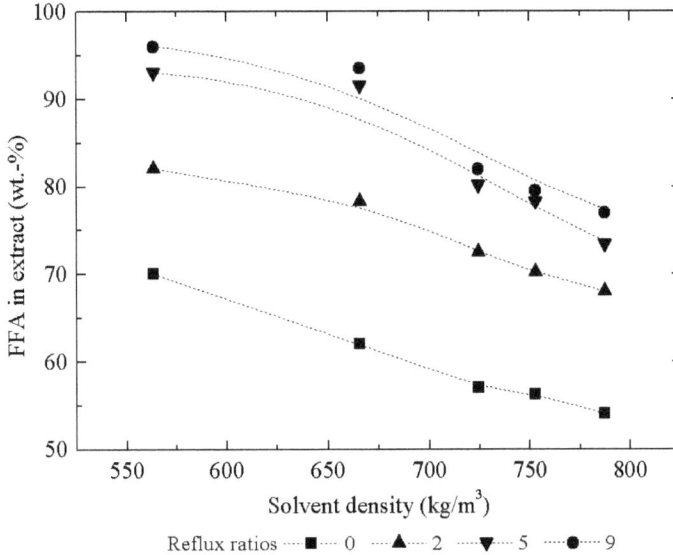

Reflux ratios ⋯■⋯ 0 ⋯▲⋯ 2 ⋯▼⋯ 5 ⋯●⋯ 9

Figure 7.20. Effect of reflux ratio on extract composition at 60 °C.

Figure 7.21 shows the influence of SFR on the extract compositions at 60 °C and 20 MPa. Since the extract sample amounts collected were in the range of some grams per hour, the accuracy of the measurements was influenced especially by experimental skills. For SFR higher than 150, the presented results were calculated as discussed in Chapter 2.3. A Microsoft Excel$^©$ 2000 flowsheet was developed for the calculations within a countercurrent fractionation column. Higher pressures were responsible for increases in the solvent loadings. As a consequence, higher reflux ratios were employed when SFR was mantained constant. Thus, a compromiss between increasing reflux ratio and decreasing selectivity could be found, balancing both effects.

For the evaluation of the concentrations of extract and raffinate samples collected, the next set of diagrams is presented. Since the refining of RBO produced larger amounts of raffinate than extract, evaluation of raffinate compositions consisted on a very important parameter for the design of a reliable fractionation process.

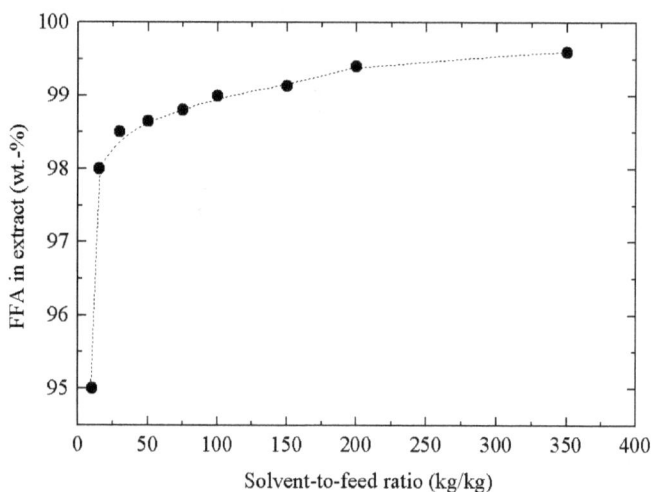

Figure 7.21. Influence of solvent-to-feed ratio on extract composition at 60 °C.

Figure 7.22 illustrates the FFA concentration profiles obtained at 60 and 80 °C, with pressures ranging from 14 to 25 MPa. Starting from a feed concentration of 10 wt.-% FFA (see Table 7.2), FFA concentrations in the extract samples varied from 70 to 93 wt.-%. After steady state was reached, these experiments were carried out during 7 hours. FFA concentrations in raffinate samples were not detected, proving that the removal of FFA from crude RBO was successfully achieved. Although the effect of temperature did not present a regular trend, the results obtained were in good agreement with the ones previously presented in the literature for the deacidification of RBO [54, 70, 99-101, 167] and other vegetable oils [102]: at lower CO$_2$ densities (563.5 and 384.02 kg/m^3 at 14 MPa/60 °C and 14 MPa/80 °C, respectively), the highest FFA concentrations were achieved. As expected, TG losses in the extract were low (concentration decreased up to approx. 7 wt.-%), due to the higher solubility as well as higher selectivity for FFA in supercritical CO$_2$ under the investigated conditions.

The TG concentrations obtained in the raffinate samples are presented in Figure 7.23. TG concentration in the crude RBO was 84.66 wt.-% and is represented in the Figure by means of the scale break. As already discussed above, TG losses in extract samples were low at the lowest pressure levels, what led to higher TG concentrations in the raffinate. The temperature effect presented higher TG concentrations at the highest temperature investigated, what was also observed by Dunford and King [54].

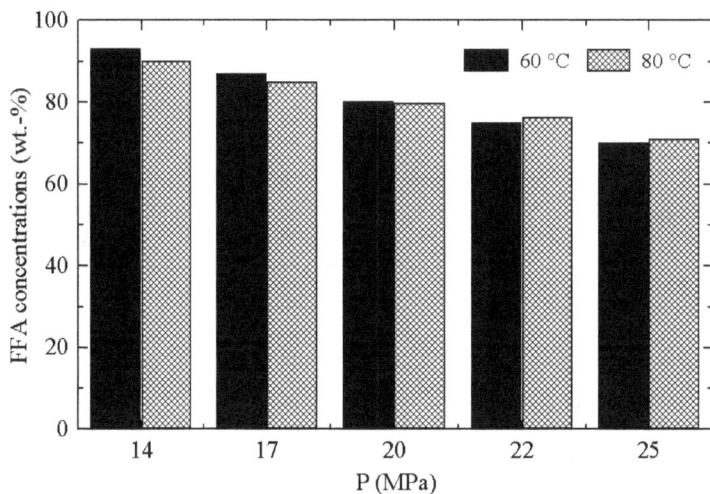

Figure 7.22. Effect of temperature and pressure on the FFA content of extract fractions.

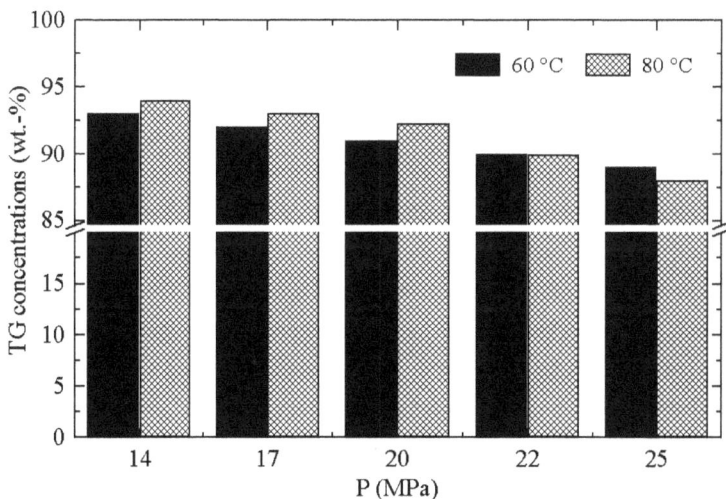

Figure 7.23. TG concentrations of the raffinate fractions at 60 and 80 °C.

Figure 7.24 presents the sterol concentration profiles obtained during the RBO fractionation. Higher sterol contents were obtained in the raffinate samples at the highest pressures employed. Additionally, the highest sterol concentrations were obtained at the lowest temperature, achieving up to 1.05 wt.-%, what corresponded to a significant enrichment, since the feed material was composed by 0.48 wt.-% sterols. These results followed the trends observed during the fractionation of palm oil derivates [145]. On the other hand, Dunford and King [54] have found significant sterol losses in the extract fractions, what was not observed in this work.

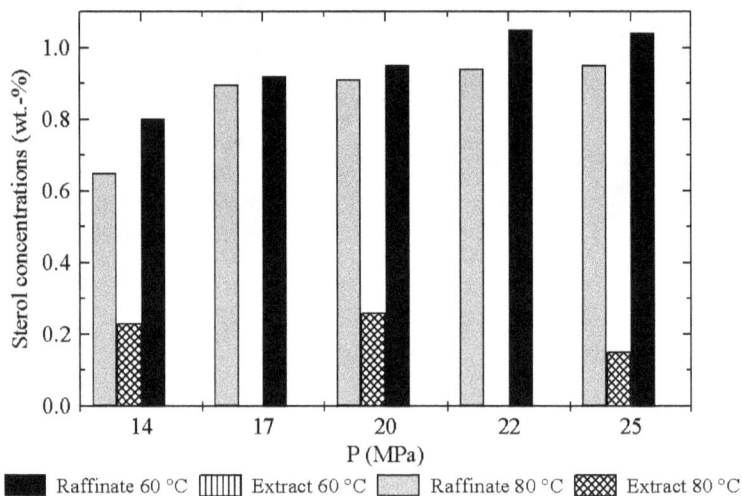

Figure 7.24. Sterol contents obtained at different operational conditions.

Oryzanol contents of the raffinate samples were significantly higher (up to 10 fold at 20 MPa) than that of the extract fractions obtained at the same conditions, as presented in Figure 7.25. The oryzanol content of the extract fractions, which ones corresponded to extract losses, did not show a clear trend with temperature and pressure, what was also observed by Dunford and King [54]. On the other hand, higher concentrations of oryzanol in the raffinate samples were achieved at higher pressure and low temperature (60 °C) levels. The results here obtained were quite interesting, since raffinate samples obtained contained higher oryzanol concentrations than a commercial oryzanol enriched RBO (0.6 wt.-%) [54], what confirmed that supercritical fractionation can be used as an alternative tool for the deacidification of RBO with a consequent enrichment of important low volatile components in the raffinate fractions.

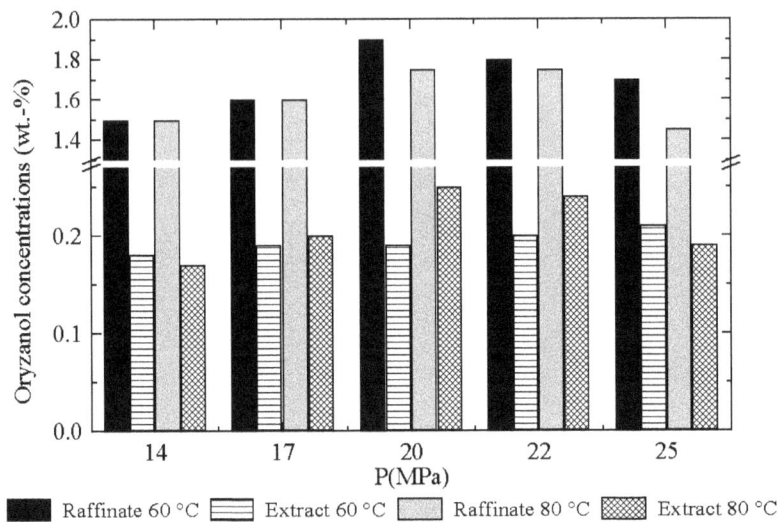

Figure 7.25. Concentration profiles of oryzanols at different operational conditions.

8. Conclusions and Outlook

8.1. Mandarin Peel Oil

In this work, the deterpenation of two mandarin peel oils of different compositions through countercurrent fractionation and ad-/desorption in silica gel with supercritical carbon dioxide were performed. The idea was to couple both processes and investigate the feasibility of each one. After performing the sets of experimets proposed, the removal of terpenes was considered successfull.

Countercurrent experiments were performed in conditions varying from 8.5 to 11.0 MPa and from 50 to 70 °C. The efficiency of the separations was confirmed through the analysis of the selectivities obtained between the most representative terpene components and the major components present in the aroma fraction. That showed that a separation of terpenes and aroma components through countercurrent fractionation under supercritical conditions was possible. For the design of a countercurrent process, theoretical selectivities and the height of equilibrium stages depended on the model used for the description of the behavior of the separation factor. Since the aroma concentrations in the feed mixtures were low in comparison with other works [112], experiments were carried out with linalool-enriched feed samples. Thus, the separation efficiency was slightly improved. At high terpene concentrations, the separation factors depended mostly on the composition of the aroma fraction. With higher linalool concentrations, higher separation factors were obtained. The process was evaluated as a good alternative to vacuum distillation.

The second set of investigations with mandarin peel oil consisted of ad-/desorption experiments. The deterpenation and fractionation of the high added value aroma components were achieved at optimal process conditions, namely 40 °C and 8.0 MPa for the terpene desorption and 20.0 MPa for the selective desorption of aroma components. A loading of 25 wt.-% was found to be the the best option, whereas good results were also obtained when using a loading of 50 wt.-%. Additionally, scale-up experiments were performed and the results have shown that an aroma fraction of high purity similar to the ones obtained in laboratory scale could be achieved.

8.2. Rice Bran Oil

Extraction and fractionation experiments with rice bran oil were conducted at different operational conditions, varying from 10 to 40 MPa and from 40 to 60 °C. Batch extractions have been also modeled considering the extraction of more than one component: 4 classes of components have been chosen, namely FFA, TG, sterols and oryzanols. The goal was to develop simulations of the extraction and purification processes in order to validate the experimental results. The results have shown that the fractionation could be achieved, especially when using higher solvent selectivities.

In order to perform the optimization of the operational conditions, phase equilibria measurements obtained were used as a basis for the separation analysis between FFA and TG. The chosen experimental parameters were pressure and temperature, reflux and solvent-to-feed ratios, and the number of separation stages. Based on these experimental data, the design of a production scale countercurrent column was possible by using a pilot scale column with an effective separation height of 6 m. The refining of crude RBO was successfully achieved: extract fractions were enriched up to approx. 95 wt.-% FFA and the concentration in the raffinate fractions was below 1 wt.-%. Besides the good results obtained for the removal of FFA (deacidification), very interesting results were obtained for sterols and oryzanols, providing substantial enrichments of these nutraceutical components in the raffinate fractions. In order to achieve higher concentrations of sterols and oryzanols, ad-/desorption experiments could be a good alternative for their fractionation, but were not investigated in this work.

References

[1]. BRUNNER, G., *Gas Extraction*. **1994**, New York: Steinkopff.

[2]. MCHUGH, M. and KRUKONIS, V., *Supercritical fluid extraction: principles and practice*. **1993**, Stoneham.

[3]. DE LA TOUR, C., *Ann. Chim. Phys.*, **1822**, 21, 127-178.

[4]. HANNAY, J. and HOGARTH, J., *On the solubility of solids in gases*. Vol. 29. **1879**, London: The Royal Society.

[5]. WILSON, R., KEITH, P., and HAYLETT, R., Liquid propane: using in dewaxing, deasphalting, and refining heavy oils, *Ind. Eng. Chem.*, **1936**, 28, 1065.

[6]. DICKERSON, N. and MEYERS, J., Solexol fractionation of menhaden oil, *JAOCS*, **1952**, 29, 235-239.

[7]. GARCÍA-REVERTER, J., BLASCO, M., and SUBIRATS, S., *Revision on supercritical extraction industrial plants trends*, in *State of the Art Book on Supercritical Fluids*, ainia. **2004**: Valencia. p. 255-266.

[8]. BRUNNER, G., Supercritical fluids: technology and application to food processing, *Journal of Food Engineering*, **2005**, 67, 21-33.

[9]. DEL VALLE, J.M., DE LA FUENTE, J.C., and CARDARELLI, D.A., Contributions to supercritical extraction of vegetable substrates in Latin America, *Journal of Food Engineering*, **2005**, 67, 35-57.

[10]. REVERCHON, E., Supercritical fluid extraction and fractionation of essential oils and related products, *Journal of Supercritical Fluids*, **1997**, 10, 1-37.

[11]. REVERCHON, E. and DE MARCO, I., Supercritical fluid extraction and fractionation of natural matter, *Journal of Supercritical Fluids*, **2006**, 38, 146-166.

[12]. ROSA, P.T.V. and MEIRELES, M.A.A., Supercritical technology in Brazil: system investigated (1994–2003), *Journal of Supercritical Fluids*, **2005**, 34, 109-117.

[13]. STAHL, E., QUIRIN, K.W., GLATZ, A., GERARD, D., and RAU, G., New developments in the field of high-pressure extraction of natural products with dense gases, *Ber. Bunsenges. Phys. Chem.*, **1984**, 88, 900-907.

[14]. SANDLER, S.I., *Chemical and engineering thermodynamics*. **1989**, New York: John Wiley and Sons.

[15]. VAN KONYNENBURG, P. and SCOTT, R., Critical lines and phase equilibria in binary van der Waals mixtures, *Phil. Trans. Roy. Soc.*, **1980**, 298, 728-730.

[16]. TREYBAL, E.T., *Mass-transfer operations*. 3rd ed. **1980**, New York: McGraw-Hill.

[17]. SATTLER, K., *Thermische Trennverfahren: Grundlagen, Auslegung, Apparate*. **1995**, Weinheim: VCH.

[18]. ROUQUEROL, F., ROUQUEROL, J., and SING, K., *Adsorption by powders and porous solids: Principles, methodology and applications*. **1999**, London: Academic Press.

[19]. SLEJKO, F.L., *Adsorption technology*. **1985**, New York: Marcel Dekker Inc.

[20]. WEDLER, G., *Adsorption*. **1970**: Verlag Chemie.

[21]. SUBRA, P., VEGA-BANCEL, A., and REVERCHON, E., Breakthrough curves and adsorption isotherms of terpene mixtures in supercritical carbon dioxide, *Journal of Supercritical Fluids*, **1998**, 12, 43-57.

[22]. MARTÍNEZ, J., MONTEIRO, A.R., ROSA, P.T.V., MARQUES, M.O.M., and MEIRELES, M.A.A., Multicomponent model to describe extraction of ginger oleoresin with supercritical carbon dioxide, *Ind. Eng. Chem. Res.*, **2003**, 42, 1057-1063.

[23]. SOVOVÁ, H., Rate of the vegetable oil extraction with supercritical CO_2 – I. Modelling of extraction curves, *Chem. Eng. Sci.*, **1994**, 49, 409-414.

[24]. SOVOVÁ, H., Mathematical model for supercritical fluid extraction of natural products and extraction curve evaluation, *Journal of Supercritical Fluids*, **2005**, 33, 35-52.

[25]. TAN, C.S. and LIU, D.C., Desorption of ethyl acetate from activated carbon by supercritical carbon dioxide, *Ind. Eng. Chem. Res.*, **1988**, 27, 988-991.

[26]. DEL VALLE, J.M. and DE LA FUENTE, J., Supercritical CO_2 extraction of oilseeds: review of kinetic and equilibrium models, *Critical Reviews in Food Science and Nutrition*, **2006**, 46(2), 131-160.

[27]. AKGERMAN, A. and MADRAS, G., *Fundamentals of solids extraction by supercritical fluids*, in *Supercritical Fluids: Fundamentals for Application*, E. Kiran and J.M.H. Levelt Sengers. **1994**, Kluwer: Dordrecht. p. 669-695.

[28]. CRANK, J., *The mathematics of diffusion*. **1975**, Oxford: Clarendon. 89-103.

[29]. BARTLE, K.D., CLIFFORD, A.A., HAWTHORNE, S.B., LANGENFELD, J.J., MILLER, D.J., and ROBINSON, R., A model for dynamic extraction using a supercritical fluid, *Journal of Supercritical Fluids*, **1990**, 3, 143-149.

[30]. CHASSAGNEZ-MÉNDEZ, A.L., CORREA, N.C.F., FRANÇA, L.F., MACHADO, N.T., and ARAÚJO, M.E., A mass transfer model applied to the supercritical extraction with CO_2 of curcumins from turmeric rhizomes (*Curcuma longa* L.), *Braz. J. Chem. Eng.*, **2000**, 17, 315-322.

[31]. CAMPOS, L.M.A.S., MICHIELIN, E.M.Z., DANIELSKI, L., and FERREIRA, S.R.S., Experimental data and modeling the supercritical fluid extraction of marigold (*Calendula officinalis*) oleoresin, *Journal of Supercritical Fluids*, **2005**, 34, 163-170.

[32]. GOTO, M., ROY, B.C., and HIROSE, T., Shrinking core leaching model for supercritical fluid extraction, *Journal of Supercritical Fluids*, **1996**, 9, 128-133.

[33]. GOTO, M., ROY, B.C., KODAMA, A., and HIROSE, T., Modeling supercritical fluid extraction process involving solute-solid interaction, *Journal of Chemical Engineering of Japan*, **1998**, 31(2), 171-177.

[34]. SPRICIGO, C.B., PINTO, L.T., BOLZAN, A., and NOVAIS, A.F., Extraction of essential oil and lipids from nutmeg by liquid carbon dioxide, *Journal of Supercritical Fluids*, **1999**, 15(3), 253-259.

[35]. KIM, K.H. and HONG, J., Equilibrium solubilities of spearmint oil components in supercritical carbon dioxide, *Fluid Phase Equilibria*, **1999**, 164, 107-115.

[36]. FAN, M. and WANG, K., Optimal harvesting policy for single population with periodic coefficients, *Math. Biosci.*, **1998**, 152, 165.

[37]. ZWIEFELHOFER, U., *Stoffübertragung in Festbetten bei der Extraktion mit überkritischem Kohlendioxid am Beispiel der Extraktion von Theobromin aus Kakaoschalen*, **1994**, PhD Thesis, Technische Universität Hamburg-Harburg, Hamburg.

[38]. ANONYMOUS, www.fao.org, **2006**.

[39]. SAUNDERS, R.M., Rice bran: composition and potential food uses, *Food Rev. Inter.*, **1985**, 86(1(3)), 465-495.

[40]. KREBS, A., Reis und Reisprodukte für die Lebensmittelindustrie, *LVT Lebensmittel Industrie*, **2000**, 45, 179-181.

[41]. JULIANO, O.B., *Rice: chemistry and technology*. **1994**, St. Paul: The American Association of Cereal Chemists Inc.

[42]. SHIN, T.S. and GODBER, J.S., Changes of endogenous antioxidants and fatty acid composition in irradiated rice bran during storage, *J. Agric. Food Chem.*, **1996**, 44(2), 567-573.

[43]. SHIN, T.S., GODBER, J.S., MARTIN, D.E., and WELLS, J.H., Hydrolytic stability and changes in E vitamers and oryzanol of extruded rice bran during storage, *J. Food Sci.*, **1997**, 62(4), 704-708.

[44]. DA SILVA, M.A., SANCHES, C., and AMANTE, E.R., Prevention of hydrolytic rancidity in rice bran, *Journal of Food Engineering*, **2006**, 75(4), 487-491.

[45]. MCPEAK, D.L., *Rice bran processing apparatus*, **1988**.

[46]. LAKKAKULA, N.R., LIMA, M., and WALKER, T., Rice bran stabilization and rice bran oil extraction using ohmic heating, *Bioresource Technology*, **2004**, 92, 157-161.

[47]. PRAKASH, J., Rice bran proteins: properties and food uses, *Critical Review of Food Science Nutrition*, **1996**, 36(6), 537-552.

[48]. PRAKASH, J. and RAMANATHAM, G., Physico-chemical and nutritional traits of rice bran protein concentrate-base weaning foods, *Journal of Food Science and Technology*, **1995**, 32(5), 395-399.

[49]. RAMEZANZADEH, F.M., RAO, R.M., WINDHAUSER, M., PRINYAWIWATKUL, R.T., and MARSHALL, W.E., Prevention of hydrolytic rancidity in bran during storage, *Journal of Agricultural and Food Chemistry*, **1999**, 47, 3050-3052.

[50]. GOPALA KRISHNA, A.G., KHATOON, S., SHIELA, P.M., SARMANDAL, C.V., INDIRA, T.N., and MISHRA, A., Effect of refining of crude rice bran oil on the retention of oryzanol in the refined oil, *JAOCS*, **2001**, 78, 127-131.

[51]. SUGANO, M. and TSUJI, E., Rice bran oil and cholesterol metabolism, *Journal of Nutrition*, **1997**, 127(3), 521-524.

[52]. ZHAO, W., SHISHIKURA, A., FUJIMOTO, K., ARAI, K., and SAITO, S., Fractional extraction of rice bran oil with supercritical carbon dioxide, *Agric. Biol. Chem.*, **1987**, 51, 1773-1777.

[53]. ANONYMOUS, http://www.ricebranoil.info/articles/frying.html, **2006**.

[54]. DUNFORD, N.T. and KING, J.W., Phytosterol enrichment of rice bran oil by a supercritical carbon dioxide fractionation technique, *Journal of Food Science*, **2000**, 65(8), 1395-1399.

[55]. ZULLAIKAH, S., LAI, C.C., VALI, S.R., and JU, Y.H., A two-step acid catalyzed process for the production of biodiesel from rice bran oil, *Bioresource Technology*, **2005**, 96, 1889-1896.

[56]. ZHENG, L., ZHENG, P., SUN, Z., BAI, Y., WANG, J., and GUO, X., Production of vanillin from waste residue of rice bran oil by *Aspergillus niger* and *Pycnoporus cinnabarinus*, *Bioresource Technology*, **2007**, 98, 1115-1119.

[57]. LONCIN, M., Über das Palmöl, *Fette Seifen Anstrichmittel*, **1974**, 1, 104-112.

[58]. RODRIGUES, C.E.C., PESSOA FILHO, P.A., and MEIRELES, A.J.A., Phase equilibrium for the system rice bran oil+fatty acids+ethanol+water+g-oryzanol+tocols, *Fluid Phase Equilibria*, **2004**, 216, 271-283.

[59]. GARCÍA, A., LUCAS, A., RINCÓN, J., ALVAREZ, A., GRACIA, I., and GARCÍA, M.A., Supercritical carbon dioxide extraction of fatty and waxy material from rice bran, *JAOCS*, **1996**, 73(9), 1127-1131.

[60]. WILSON, T.A., AUSMAN, L.M., LAWTON, C.W., HEGSTED, D.M., and NICOLOSI, R.J., Comparative cholesterol lowering properties of vegetable oils: beyond fatty acids, *Journal of the American College of Nutrition*, **2000**, 19(5), 601-607.

[61]. CICERO, A.F.G. and GADDI, A., Rice bran oil and γ-oryzanol in the treatment of hyperlipoproteinaemias and other conditions, *Phytother. Res.* , **2001**, 15, 277-289.

[62]. ORTHOEFER, F., Rice bran oil: health lipid source, *Food Technology*, **1996**, 50(12), 62-64.

[63]. CARROL, L.E., Functional properties and applications of stabilized rice in bakery products, *Food Technology*, **1990**, 74-76.

[64]. MACHADO, N.T., *Fractionation of PFAD-compounds in countercurrent columns using supercritical carbon dioxide as solvent*, **1998**, PhD Thesis, Technische Universität Hamburg-Harburg, Hamburg.

[65]. MCCASKILL, D.R. and ZHANG, F., Use of rice bran oil in foods, *Food Technology*, **1999**, 53(2), 50-53.

[66]. SINGH, B., SEKHON, K.S., and SINGH, N., Suitability of full fat and defatted rice bran obtained from Indian rice for use in food products, *Plant Foods for Human Nutrition*, **1995**, 47, 191-200.

[67]. ABIDI, S.L. and RENNICK, K.A., Determination of nonvolatile components in polar fractions of rice bran oils, *JAOCS*, **2003**, 80(1), 1057-1062.

[68]. BOT, A. and AGTEROF, G.M., Structuring of edible oils by mixtures of γ-oryzanol with β-sitosterol or related phytosterols, *JAOCS*, **2006**, 83(6), 513-521.

[69]. VISSERS, M.N., ZOCK, P.L., MEIJER, G.W., and KATAN, M.B., Effect of plant sterols from rice bran oil and triterpene alcohols from sheanut oil on serum lipoprotein concentrations in humans, *Am. J. Clin. Nutr.*, **2000**, 72, 1510-1515.

[70]. DUNFORD, N.T., TEEL, J.A., and KING, J.W., A continuous countercurrent supercritical fluid deacidification process for phytosterol ester fortification in rice bran oil, *Food Research International*, **2003**, 36, 175-181.

[71]. HEMAVATHY, J. and PRABHAKAR, J.V., Lipid composition of rice (*Oryza sativa* L.) bran, *JAOCS*, **1987**, 64(7), 1016-1019.

[72]. JIANG, Y. and WANG, T., Phytosterol in cereal by-products, *JAOCS*, **2005**, 82(6), 439-444.

[73]. KANEKO, R. and TSUCHIYA, T., New compound in rice bran and germ oil, *Kogyo Kagaku Zasshi*, **1954**, 57, 526.

[74]. MILLER, A., *Analytik von Minorlipiden: Ferulasäureester von Phytosterolen (γ-Oryzanol) in Reis*, **2004**, PhD Thesis, Technische Universität München, Munich.

[75]. JAHN, G.B., *Extração supercrítica do óleo de farelo de arroz e obtenção de frações enriquecidas em γ-oryzanol*, **2004**, Master Thesis, Universidade Federal de Santa Catarina, Florianópolis, Brazil.

[76]. KAIMAL, T.N.B., γ-Oryzanol from rice bran oil, *J. Oil Technol. Assoc. India*, **1999**, 31, 83-91.

[77]. LLOYD, B.J., SIEBENMORGEN, T.J., and BEERS, K.W., Effects of commercial processing on antioxidants in rice bran, *Cereal Chem.*, **2000**, 77(5), 551-555.

[78]. JULIANO, C., COSSU, M., ALAMANNI, M.C., and PIU, L., Antioxidant activity of gamma-oryzanol: Mechanism of action and its effect on oxidative stability of pharmaceutical oils, *International Journal of Pharmaceutics*, **2005**, 299, 146-154.

[79]. KAISER, B., *Ein Sonnenschutzmittel für Haut und Haar mit verbesserter Schutzeigenschaft*, **1995**, Deutsche Patentanmeldung DE 4421038 A1.

[80]. KAISER, B., *Sonnenschutzmittel für die topische Anwendung am menschlichen Körper*, **1997**, Deutsche Patentanmeldung DE 4421038 C2.

[81]. BHOSLE, B.M. and SUBRAMANIAN, R., New approaches in deacidification of edible oils – a review, *Journal of Food Engineering*, **2005**, 69, 481-494.

[82]. ČMOLÍK, J. and POKORNÝ, J., Physical refining of edible oils, *Eur. J. Lipid Sci. Technol.*, **2000**, 102, 472-486.

[83]. MAMIDIPALLY, P.K. and LIU, S.X., First approach on rice bran oil extraction using limonene, *Eur. J. Lipid Sci. Technol.*, **2004**, 106, 122-125.

[84]. MONSOOR, M.A., PROCTOR, A., and HOWARD, L.R., Aqueous extraction, composition, and functional properties of rice bran emulsion, *JAOCS*, **2003**, 80, 361-365.

[85]. MONSOOR, M.A. and PROCTOR, A., Tocopherol, tocotrienol and oryzanol content of rice bran aqueous extracts, *JAOCS*, **2005**, 82(6), 463-464.

[86]. PROCTOR, A. and BOWEN, D.J., Ambient-temperature extraction of rice bran oil with hexane and isopropanol, *JAOCS*, **1996**, 73(6), 811-813.

[87]. TANAKA, T., HOSHINA, M., TANABE, S., SAKAI, K., OHTSUBO, S., and TANIGUCHI, M., Production of D-lactic acid from defatted rice bran by simultaneous saccharification and fermentation, *Bioresource Technology*, **2006**, 97, 211-217.

[88]. GINGRAS, L., Refining of rice bran oil, *Inform*, **2000**, 11, 1196-1203.

[89]. NARAYAN, A.V., BARHATE, R.S., and RAGHAVARAO, K.S.M.S., Extraction and purification of oryzanol from rice bran oil and rice bran oil soapstock, *JAOCS*, **2006**, 83(8), 663-670.

[90]. ANDERSON, D., *A primer on oils processing technology.* Bailey's industrial fat and oil products, ed. Y.H. Hui. Vol. 4. **1996**, New York. 1-60.

[91]. SHEN, Z., PALMER, M.P., TING, S.S.T., and FAIRCLOUGH, R.J., Pilot scale extraction of rice bran oil with dense carbon dioxide, *J. Agric. Food Chem.*, **1996**, 44, 3033-3039.

[92]. MAHESHWARI, P., NIKOLOV, Z.L., WHITE, T.M., and HARTEL, R., Solubility of fatty acids in supercritical carbon dioxide, *JAOCS*, **1992**, 69(11), 1069-1076.

[93]. SHEN, Z., PALMER, M.P., TING, S.S.T., and FAIRCLOUGH, R.J., Pilot scale extraction and fractionation of rice bran oil using supercritical carbon dioxide, *J. Agric. Food Chem.*, **1997**, 45, 4540-4544.

[94]. KUK, M.S. and DOWD, M.K., Supercritical CO_2 extraction of rice bran, *JAOCS*, **1998**, 75(5), 623-628.

[95]. KIM, H.J., LEE, S.B., PARK, K.A., and HONG, I.K., Characterization of extraction and separation of rice bran oil rich in EFA using SFE process, *Separation and Purification Technology*, **1999**, 15, 1-8.

[96]. XU, Z.M. and GODBER, J.S., Comparison of supercritical fluid and solvent extraction methods in extracting γ-oryzanol from rice bran, *JAOCS*, **2000**, 77(5), 547-551.

[97]. PERRETTI, G., MINIATI, E., MONTANARI, L., and FANTOZZI, P., Improving the value of rice by-products by SFE, *Journal of Supercritical Fluids*, **2003**, 26, 63-71.

[98]. SPARKS, D., HERNANDEZ, R., M., Z., BLACKWELL, D., and FLEMING, T., Extraction of rice bran oil using supercritical carbon dioxide and propane, *JAOCS*, **2006**, 83(10), 885-891.

[99]. DUNFORD, N.T. and KING, J.W., Thermal gradient deacidification of crude rice bran oil utilizing supercritical carbon dioxide, *JAOCS*, **2001**, 78(2), 121-125.

[100]. DUNFORD, N.T. and KING, J.W., *Supercritical fluid fractionation process for phytosterol ester enrichment vegetable oils*, **2004**, US Patent no. 6677469.

[101]. KING, J.W. and DUNFORD, N.T., Phytosterol-enriched triglyceride fractions from vegetable oil deodorizer distillates utilizing supercritical fluid fractionation technology, *Separation Science and Technology*, **2002**, 37(2), 451-462.

[102]. BRUNETTI, L., DAGHETTA, A., FEDELLI, E., KIKIC, I., and ZANDERIGHI, L., Deacidification of olive oils by supercritical carbon dioxide, *JAOCS*, **1989**, 66, 209-217.

[103]. ZIEGLER, G.R. and LIAW, Y.J., Deodorization and deacidification of edible oils with dense carbon dioxide, *JAOCS*, **1993**, 70(10), 947-953.

[104]. REEVE, D. and ARTHUR, D., Riding the citrus trail: when is a mandarin a tangerine?, *Perfumer and Flavorist*, **2002**, 27, 20-22.

[105]. WASSMANN, T., Siegeszug mit Süß und Sauer, *Kosmos*, **1992**, 1, 32-39.

[106]. TANAKA, T., *Citologia: semi centennial commemoration papers on Citrus studies* **1961**, Citologia supporting foundation: Osaka. p. 114.

[107]. MOORE, G.A., Oranges and lemons: clues to the taxonomy of Citrus from molecular markers, *Trends in Genetics*, **2001**, 17, 536-540.

[108]. MERLE, H., MORON, M., BLAZQUEZ, M.A., and BOIRA, H., Taxonomical contribution of essential oils in mandarins cultivars, *Biochemical Systematics and Ecology*, **2004**, 32, 491-497.

[109]. LOTA, M., DE ROCCA SERRA, D., TOMI, F., and CASANOVA, J., Chemical variability of peel and leaf essential oils of mandarins from *Citrus reticulata* Blanco, *Biochemical Systematics and Ecology*, **2000**, 28, 61-78.

[110]. ATTI DOS SANTOS, A., ATTI-SERAFINI, L., and CASSEL, E., Supercritical carbon dioxide extraction of mandarin (*Citrus deliciosa* tenore) from south Brazil, *Perfumer and Flavorist*, **2000**, 25, 26-36.

[111]. ATTI DOS SANTOS, A.C., ATTI-SERAFINI, L., and CASSEL, E., *Estudos de processos de extração de óleos essenciais e bioflavonóides de frutas cítricas*. **2003**, Caxias do Sul: Editora da Universidade de Caxias do Sul. 112 p.

[112]. BUDICH, M., *Countercurrent extraction of citrus aroma from aqueous and nonaqueous solutions using supercritical carbon dioxide*, **1999**, PhD Thesis, Technische Universität Hamburg-Harburg, Hamburg.

[113]. CHISHOLM, M.G., JELL, J.A., and CASS JR., D.M., Characterization of the major odorants found in the peel oil of *Citrus reticulata* Blanco cv. Clementine using gas chromatography-olfactometry, *Flavour and Fragrance Journal*, **2003**, 18, 275-281.

[114]. BLANCO TIRADO, C., STASHENKO, E.E., COMBARIZA, M.Y., and MARTINEZ, J.R., Comparative study of Colombian citrus oils by high-resolution gas chromatography and gas chromatography-mass spectrometry, *Journal of Chromatography A*, **1995**, 697, 501-513.

[115]. CHOI, H., Volatile constituents of satsuma mandarins growing in Korea, *Flavour and Fragrance Journal*, **2004**, 19, 406-412.

[116]. DUGO, P., MONDELLO, L., FAVOINO, O., CICERO, L., ZENTENO, N.A.R., and DUGO, G., Characterization of cold-pressed Mexican dancy tangerine oils, *Flavour and Fragrance Journal*, **2005**, 20, 60-66.

[117]. LOTA, M., DE ROCCA SERRA, D., TOMI, F., and CASANOVA, J., Chemical variability of peel and leaf essential oils of 15 species of mandarins, *Biochemical Systematics and Ecology*, **2001**, 29, 77-104.

[118]. NJOROGE, S.M., KOAZE, H., MWANIKI, M., TU, N.T.M., and SAWAMURA, M., Essential oils of Kenyan Citrus fruits: volatile components of two varieties of mandarins (*Citrus reticulata*) and a tangelo (*C. paradisi* × *C. tangerina*), *Flavour and Fragrance Journal*, **2005**, 20, 74-79.

[119]. VERZERA, A., MONDELLO, L., TROZZI, A., and DUGO, P., On the genuineness of citrus essential oils. Part LII. Chemical characterization of essential oil of three cultivars of *Citrus clementine* Hort, *Flavour and Fragrance Journal*, **1997**, 12, 163-172.

[120]. DUGO, P., MONDELLO, L., DUGO, L., STANCANELLI, R., and DUGO, G., LC-MS for the identification of oxygen heterocyclic compounds in citrus essential oils, *Journal of Pharmaceutical and Biomedical Analysis*, **2000**, 24, 147-154.

[121]. LAWRENCE, B.M., Progress in essential oils: mandarin oil, *Perfumer and Flavorist*, **1996**, 21, 25-28.

[122]. LAWRENCE, B.M., Progress in essential oils: mandarin oil, *Perfumer and Flavorist*, **2001**, 26, 36-44.

[123]. REVERCHON, E., Supercritical desorption of limonene and linalool from silica gel: experiments and modelling, *Chemical Engineering Science*, **1997**, 52, 1019-1027.

[124]. CULLY, J., SCHUTZ, E., and VOLBRECHT, H., *Process for separating terpenes from essential oils*, **1990**, European Union patent: EP0363971A2.

[125]. YAMAUCHI, Y. and SATO, M., Fractionation of lemon-peel oil by semi-preparative supercritical fluid chromatography, *Journal of Chromatography A*, **1990**, 505, 237-246.

[126]. ZETZL, C., *Desorption von Sauerstoffverbindungen des Zitrusöls durch überkritisches CO_2*, **1994**, Diplom Thesis, Universität Karlsruhe, Karlsruhe.

[127]. BARTH, D., CHOUCHI, D., DELLA PORTA, G., REVERCHON, E., and PERRUT, M., Desorption of lemon peel oil by supercritical carbon dioxide: deterpenation and psoralens elimination, *Journal of Supercritical Fluids*, **1994**, 7, 177-183.

[128]. DELLA PORTA, G., REVERCHON, E., CHOUCHI, D., and BARTH, D. *Citrus peel oils processing by SC-CO_2 desorption: deterpenation and high molecular weight compounds elimination.* in *Third Congresso sui Fluidi Supercritici e loro Applicazioni*. **1995**. p. 139.

[129]. GOTO, M., SATO, M., KODAMA, A., and HIROSE, T., Application of supercritical fluid technology to citrus oil processing, *Physica B: Condensed Matter*, **1997**, 239, 167-170.

[130]. SATO, M., KONDO, M., GOTO, M., KODAMA, A., and HIROSE, T., New fractionation process for citrus oil by pressure swing adsorption in supercritical carbon dioxide, *Chem. Eng. Sci.*, **1998**, 53, 4095-4104.

[131]. GOTO, M., FUKUI, G., WANG, H., KODAMA, A., and HIROSE, T., Deterpenation of bergamot oil by pressure swing adsorption in supercritical carbon dioxide, *Journal of Chemical Engineering of Japan*, **2002**, 35, 372-376.

[132]. PITOL FILHO, L., *Fracionamento de óleo de casca de laranja por dissolução em dióxido de carbono supercrítico e adsorção em silica gel*, **1999**, Master Thesis, Universidade Federal de Santa Catarina, Florianópolis.

[133]. REVERCHON, E., LAMBERTI, G., and SUBRA, P., Modelling and simulation of the supercritical adsorption of complex terpene mixtures, *Chemical Engineering Science*, **1998**, 53, 3537-3544.

[134]. ARAÚJO, J.M.A. and FARIAS, A.P.S.F., Redução do teor de limoneno e bergapteno do óleo essencial de bergamota adsorvido em sílica gel pelo CO_2 supercrítico, *Ciência e Tecnologia de Alimentos*, **2003**, 23(2), 112-115.

[135]. GERARD, D., Kontinuierliche Deterpenierung ätherischer Öle durch Gegenstromextraktion mit verdichtetem Kohlendioxid, *Chem. Ing. Tech.*, **1984**, 10, 794-795.

[136]. STAHL, E., W., Q.K., and GERARD, D., *Verdichtete Gase zur Extraktion und Raffination.* **1987**, Berlin: Springer.

[137]. SIMÕES, P.C., MATOS, H.A., CARMELO, P.J., DE AZEVEDO, E.G., and NUNES DA PONTE, M., Mass transfer in countercurrent packed columns: application to supercritical CO_2 extraction of terpenes, *Industrial & Engineering Chemistry Research*, **1995**, 34, 613-618.

[138]. REVERCHON, E., MARCIANO, A., and POLETTO, M., Fractionation of a peel oil key mixture by supercritical CO_2 in a continuous tower, *Industrial & Engineering Chemistry Research*, **1997**, 36, 4940-4948.

[139]. SATO, M., KONDO, M., GOTO, M., KODAMA, A., and HIROSE, T., Fractionation of citrus oil by supercritical countercurrent extractor with side-stream withdrawal, *Journal of Supercritical Fluids*, **1998**, 13, 311-317.

[140]. KONDO, M., GOTO, M., KODAMA, A., and HIROSE, T., Separation performance of supercritical carbon dioxide extraction columns for the citrus oil processing: observation using simulator, *Separation Science and Technology*, **2002**, 37, 3391-3406.

[141]. POIANA, M., MINCIONE, A., GIONFRIDDO, F., and CASTALDO, D., Supercritical carbon dioxide separation of bergamot essential oil by a countercurrent process, *Flavour and Fragrance Journal*, **2003**, 18, 429-435.

[142]. GIRONI, F. and MASCHIETTI, M., Supercritical carbon dioxide fractionation of lemon oil by means of a batch process with an external reflux, *Journal of Supercritical Fluids*, **2005**, 35, 227-234.

[143]. GOTTSCHAU, T., *Untersuchungen zur Anreicherung von Tocopherolen mittels Hochdruckgegenstromextraktion mit überkritischem Kohlendioxid*, **1994**, PhD Thesis, Technische Universität Hamburg-Harburg, Hamburg.

[144]. SMITH, D.C., FORLAND, S., BACHANOS, E., MATEJKA, M., and BARRETT, V., Qualitative analysis of citrus fruit extracts by GC/MS: an undergraduate experiment, *Chem. Educator*, **2001**, 6, 28-31.

[145]. GAST, K., *Enrichment of vitamin E and provitamin A from palm oil derivates with supercritical fluids*, **2006**, Technische Universität Hamburg-Harburg, Hamburg.

[146]. TUFEU, R., SUBRA, P., and PLATEAUX, C., Contribution to the experimental determination of the phase diagrams of some (carbon dioxide + a terpene) mixtures, *J. Chem. Thermodyn.*, **1993**, 25, 1219-1228.

[147]. IWAI, Y., MOROTOMI, T., SAKAMOTO, K., KOGA, Y., and ARAI, Y., High-pressure vapor-liquid equilibria for carbon dioxide+limonene, *J. Chem. Eng. Data*, **1996**, 41, 951-952.

[148]. SUZUKI, J. and NAGAHAMA, K., Measurement and correlation of solubility of limonene and linalool in high-pressure carbon dioxide, *Kagaku Kogaku Ronbunshu*, **1996**, 22, 195-200.

[149]. BERTUCCO, A., GUARISE, G.B., ZANDEGIACOMO-RIZIO, A., PALLADO, P., and VIEIRA DE MELO, S. *Binary and ternary vapor-liquid equilibrium data for the system CO_2-limonene-linalool. in 4th Italian Conference on Supercritical Fluids.* **1997**. Capri. p. 393-400.

[150]. VIEIRA DE MELO, S.A.B., *Deterpenação do óleo essencial de laranja usando CO_2 supercrítico*, **1997**, PhD Thesis, Universidade Federal do Rio de Janeiro, Rio de Janeiro, Brazil.

[151]. IWAI, Y., HOSOTANI, N., MOROTOMI, T., KOGA, Y., and ARAI, Y., High-pressure vapor-liquid equilibria for carbon dioxide+linalool, *J. Chem. Eng. Data*, **1994**, 39, 900-902.

[152]. FRANCESCHI, E., GRINGS, M.B., FRIZZO, C.D., OLIVEIRA, J.V., and DARIVA, C., Phase behavior of lemon and bergamot peel oils in supercritical CO_2, *Fluid Phase Equilibria*, **2004**, 226, 1-8.

[153]. PERRE, C., DELESTRE, G., SCHRIVE, L., and CARLES, M. *Deterpenation process for citrus oils by supercritical CO_2 extraction in a packed column. in 3rd Int. Symp. Supercritical Fluids.* **1994**. Strassbourg, France. p. 465-470.

[154]. TEMELLI, F., *Supercritical carbon dioxide extraction of terpenes from cold-pressed valencia orange oil*, **1987**, PhD Thesis, University of Florida, Gainesville, USA

[155]. STUART, G.R., DARIVA, C., and OLIVEIRA, J.V., High-pressure vapor-liquid equilibria data for CO_2-orange peel oil, *Brazilian Journal of Chemical Engineering*, **2000**, 17(2), 181-189.

[156]. RAEISSI, S. and PETERS, C.J., Experimental determination of high-pressure phase equilibria of the ternary system carbon dioxide+limonene+linalool, *Journal of Supercritical Fluids*, **2005**, 35, 10-17.

[157]. RAEISSI, S., ASENSI, J.C., and PETERS, C.J., Phase behavior of the binary system ethane+linalool, *Journal of Supercritical Fluids*, **2002**, 24 (2), 111-121.

[158]. RAEISSI, S. and PETERS, C.J., Phase behavior of the binary system ethane+limonene, *Journal of Supercritical Fluids*, **2002**, 22 (2), 93-102.

[159]. RAEISSI, S. and PETERS, C.J., Liquid-vapor and liquid-liquid-vapor equilibria of the ternary system ethane+limonene+linalool, *Journal of Supercritical Fluids*, **2005**, 33 (3), 201-208.

[160]. SAMPAIO DE SOUSA, A.R., RAEISSI, S., AGUIAR-RICARDO, A., DUARTE, C.M.M., and PETERS, C.J., High pressure phase behavior of the system ethane+orange peel oil, *Journal of Supercritical Fluids*, **2004**, 29, 59-67.

[161]. PERRY, R.H., GREEN, D.W., and MALONEY, J.O., *Chemical Engineers' Handbook.* 6th ed. **1984**, New York: McGraw-Hill.

[162]. DIAZ, S., ESPINOSA, S., and BRIGNOLE, E., Citrus peel oil deterpenation with supercritical fluids: optimal process and solvent cycle design, *Journal of Supercritical Fluids*, **2005**, 35, 49-61.

[163]. DONOSO, J.P.M., *Personal communication.* **2006**.

[164]. SCHWÄNKE, C., *Combination of countercurrent multistage extraction and ad-/desorption for deterpenation of mandarin peel oil with supercritical CO_2*, **2006**, Master Thesis, Technische Universität Hamburg-Harburg, Hamburg.

[165]. RODRIGUES, V.M., SOUSA, E.M.B.D., MONTEIRO, A.R., CHIAVONE-FILHO, O., MARQUES, M.O.M., and MEIRELES, M.A.A., Determination of the solubility of extracts from vegetable raw material in pressurized CO_2: a pseudo-ternary mixture formed by cellulosic structure+solute+solvent, *Journal of Supercritical Fluids*, **2002**, 22, 21-36.

[166]. REVERCHON, E. and IACUZIO, G., Supercritical desorption of bergamot peel oil from silica gel-Experiments and mathematical modelling, *Chemical Engineering Science*, **1997**, 52, 3553-3559.

[167]. KING, J.W., DEMESSIE, E.S., TEMELLI, F., and TEEL, J.A., Thermal gradient fractionation of glyceride mixtures under supercritical fluid conditions, *Journal of Supercritical Fluids*, **1997**, 10, 127-137.

[168]. KALE, V., KATIKANENI, S.P.R., and CHERYAN, M., Deacidifying rice bran oil by solvent extraction and membrane technology, *JAOCS*, **1999**, 76(6), 723-727.

[169]. BAMBERGER, T., ERICKSON, J.C., and COONEY, C.L., Measurement and model prediction of solubilities of pure fatty acids, pure triglycerides, and mixtures of triglycerides in supercritical carbon dioxide *J. Chem. Eng. Data*, **1988**, 33, 327-333.

[170]. JUNGFER, M., *Gegenstromtrennung von schwerflüchtigen Naturstoffen mit überkritischen komprimierten Gasen unter Verwendung von Schleppmitteln*, **2000**, PhD Thesis, Technische Universität Hamburg-Harburg, Hamburg.

Appendix A : Response Factors for GC

Table A.1. Response factors for the standard components used in the RBO GC analysis.

Component	M (g/mol)	Purity (%)	Origin	Retention time (min)	RF (-)
FFA					
Myristic acid	228.4	98.0	Merck[a]	6.2	1.35
Palmitic acid	256.4	99.0	Merck[a]	7.5	1.48
Oleic acid	282.4	98.0	Merck[a]	8.6	1.49
Linoleic acid	280.4	99.0	Merck[a]	8.8	1.52
Sterols					
Campesterol	400.7	40.0	Sigma[b]	19.9	1.25
Stigmasterol	412.7	96.0	Sigma[b]	20.3	1.05
β-sitosterol	414.7	60.0	Sigma[b]	21.4	1.11
Triglycerides					
Tripalmitin	807.3	99.0	Sigma[b]	31.2	1.42
Triolein	885.3	99.0	Sigma[b]	33.0	1.65
Tristearin	891.5	99.0	Sigma[b]	33.2	1.78
Internal standard					
Squalan	422.8	97.0	Merck[a]	12.5	-

[a] – Merck KGaA, Darmstadt, Germany.
[b] – Sigma Aldrich Chemie GmbH, Taufkirchen, Germany.

Appendix B : Experimental Results for Mandarin Peel Oil

GC-MS analysis

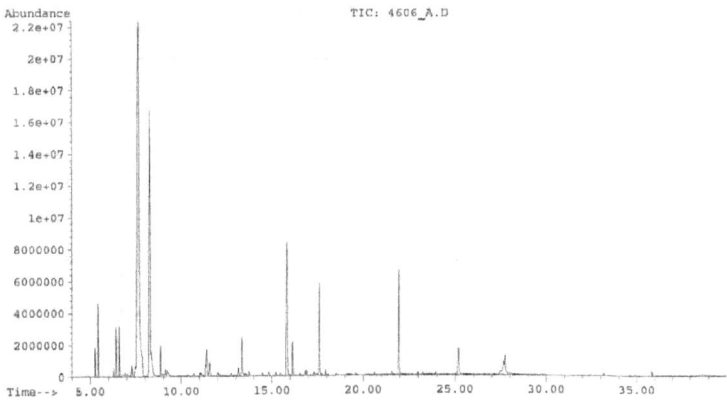

Figure B.1. GC-MS analysis obtained for crude Spanish MPO.

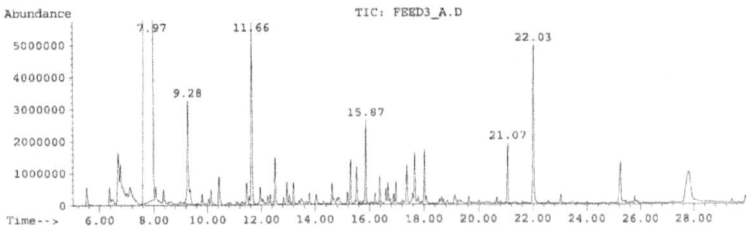

Figure B.2. GC-MS analysis obtained for crude Brazilian MPO.

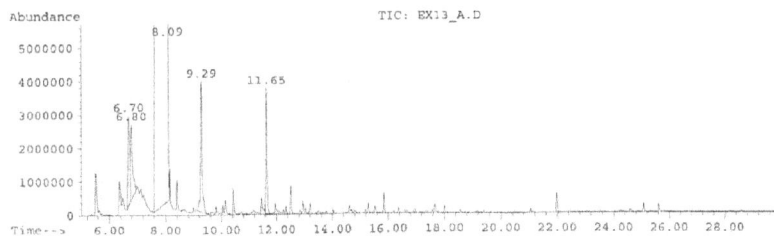

Figure B.3. GC-MS analysis obtained for an extract sample (Spanish red MPO).

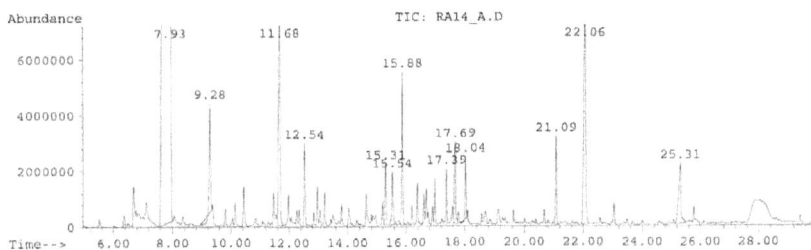

Figure B.4. GC-MS analysis obtained for a raffinate sample (Spanish red MPO).

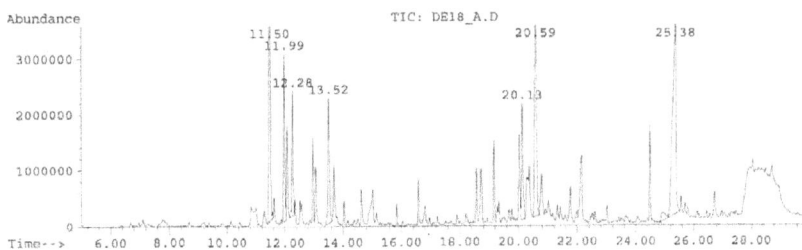

Figure B.5. GC-MS analysis obtained for sample obtained during a desorption experiment (Spanish red MPO) at 20 MPa, 40 °C after 3 hours.

Table B.1. Qualitative composition of mandarin peel oil obtained by GC-MS.

No.	identified substance	red	green	No.	identified substance	red	green	No.	identified substance	red	green
1	α-Thujene	a	x	21	**Decanal**	**x**	**x**	41	α-Carophyllene	x	x
2	α-Pinene	x	x	22	Carvone	-	x	42	α-Farnesene	x	-
3	Methyl-heptanon	-	d	23	**Citral isomer**	**x**	**x**	43	α-Bergamotene	-	x
4	Camphene	-	d	24	Geraniol	b	-	44	Bisabulene	x	-
5	Sabinene	x	x	25	**Citral isomer**	**x**	**x**	45	Cadina-3,9-diene	x	x
6	Methylheptenon	b	c	26	Decenal	x	-	46	Elemol	x	x
7	**β-Myrcene**	**x**	**x**	27	Trimethyl-cyclohexenon	-	-	47	Dodecanoic acid	x	x
8	β-Myrcene isomer	x	-	28	Perillaldehyde	x	x	48	Phthalic acid ester	-	d
9	**Octanal**	**x**	-	29	Thymol	x	x	49	Spathulenol	b	-
10	α-Terpinene	-	x	30	Methyl-Anthranilte	-	x	50	Ledol	b	-
11	**d-Limonene**	**x**	**x**	31	Decadienal	x	-	51	Octahydrotetramethyl-naphtalinmethanol	b	d
12	Dimethyloctatriene	x	-	32	Neryl-/Geranyl-Acetat	x	-	52	2,6,10-Trimethyl-2,6,9,11-decatetraene	x	-
13	**γ-Terpinene**	-	x	33	Decanoic acid	b	x	53	Trimethyl-dodecatrienal	b	-
14	1-Isobutyl-4-methyl-cyclohexene	x	-	34	Ylangene	-	x	54	Sinensal	x	x
15	Terpinolene	-	x	35	Elemene	-	x	55	Tetradecanoic acid	b	-
16	**Linalool**	**x**	**x**	36	Cububen	x	-	56	Trimethyl-pentadecatrienon	b	-
17	Nonanal	b	-	37	Dodecanal	x	-	57	Hexadecanoic acid	x	x
18	Citronellal	x	-	38	Methyl-N-methyl-Anthranilate	-	x	58	Phytol	-	x
19	**Terpineol (different isomers)**	**x**	**x**	39	Caryophyllene	x	x	59	Octadecanoic acid	x	x
20	**Octanoic acid**	**x**	**x**	40	Farnesene isomer	x	-				

bold: *confirmed by standard* a) *identified only in extract of countercurrent extraction,* b) *identified only in sample of supercritical desorption at 200 bar,* c) *identified only in raffinate of countercurrent extraction,* d) *identified only in desorption residue recovered with ethanol*

Table B.2. Experimental conditions investigated in the countercurrent fractionation of MPO.

Experiment number	P (MPa)	T (°C)	CO_2 (kg/h)	Feed (g/h)
1	8.0	50	2.5	145
2	8.5	50	2.0	57
3	8.5	50	3.0	58
4	11.0	70	1.8	126
5	11.5	70	1.9	130
6	8.5	50	3.2	95
7	8.7	50	4.0	100
8	9.0	50	3.9	98
9	10.0	60	2.2	43
10	10.0	60	2.3	41
11	10.0	60	2.1	41
12	9.0	50	1.9	67
13	9.0	50	3.1	71
14	10.0	60	1.9	106
15	10.0	60	1.9	57
16	9.0	50	3.0	73

Table B.3. Average composition of samples from countercurrent extraction of red Spanish oil without reflux.

Experiment	1	2	3	4	5
Extract					
α-Pinene	0.58	0.62	0.61	0.62	0.65
Sabinene	0.64	0.67	0.67	0.67	0.71
Myrcene	1.98	1.93	1.99	2.00	2.08
Limonene	96.44	96.53	96.42	95.05	96.17
Linalool	0.24	0.18	0.21	0.18	0.20
Rest	0.13	0.07	0.09	1.49	0.19
Raffinate					
α-Pinene	0.46	0.44	0.49	0.44	0.36
Sabinene	0.55	0.55	0.58	0.55	0.48
Myrcene	1.74	1.66	1.73	1.86	1.41
Limonene	96.50	96.65	96.43	96.44	95.95
Linalool	0.27	0.26	0.28	0.24	0.24
Decanal	0.23	0.19	0.21	0.11	0.21
other aroma	-	-	-	-	0.06
Sinensal	0.19	0.21	0.20	0.18	0.30
Rest	0.07	0.05	0.07	0.20	0.99

Table B.4. Average composition of samples from countercurrent extraction of red Spanish oil with reflux.

Experiment	6	7	8	9	10	11
Extract						
α-Pinene	0.60	0.68	0.58	0.75	0.71	0.78
Sabinene	0.66	0.72	0.64	0.74	0.75	0.80
Myrcene	1.86	2.10	1.83	2.19	2.17	2.34
Limonene	93.75	96.18	91.32	95.80	96.20	95.41
Linalool	0.18	0.15	0.23	0.18	0.11	0.62
Rest	2.95	0.18	5.40	0.34	0.05	0.06
Raffinate						
α-Pinene	0.45	0.44	0.26	0.29	0.19	0.27
Sabinene	0.55	0.54	0.43	0.48	0.33	0.41
Myrcene	1.64	1.70	1.43	1.36	1.11	1.31
Limonene	95.77	96.37	95.16	95.76	96.87	93.56
Linalool	0.27	0.31	0.33	0.27	0.31	3.98
Decanal	0.19	0.25	0.40	0.22	0.27	0.19
other aroma	-	-	-	-	0.13	0.03
Sinensal	0.19	0.22	0.36	0.07	0.31	0.10
Rest	0.93	0.19	1.63	1.55	0.47	0.14

Table B.5. Average composition of samples from countercurrent extraction of green Brazilian oil with reflux.

Experiment	12	13	14	15	16
Extract					
α-Thujene	0.76	0.78	0.83	1.03	0.81
α-Pinene	2.21	2.32	2.24	2.80	2.36
Sabinene	1.94	1.97	1.89	2.20	1.99
Myrcene	1.85	1.87	1.84	2.13	1.87
Limonene	70.96	70.69	69.37	71.04	70.15
γ-Terpinene	19.70	19.54	19.03	18.55	19.40
Terpinolene	0.87	0.85	0.85	0.72	0.85
Linalool	0.13	0.12	0.15	0.09	0.93
Other aroma	0.43	0.44	0.57	0.05	0.45
Sinensal	-	-	0.15	-	-
Rest	1.15	1.42	3.10	1.40	1.20
Raffinate					
α-Thujene	-	-	0.33	0.18	-
α-Pinene	0.86	0.61	1.10	0.64	0.53
Sabinene	0.24	0.21	1.35	1.19	0.34
Myrcene	0.17	0.48	1.29	1.14	0.61
Limonene	31.51	64.87	67.63	65.11	60.88
γ-Terpinene	19.34	21.70	20.59	21.04	20.72
Terpinolene	1.91	1.32	1.01	1.09	1.20
Linalool	0.48	0.25	0.16	0.17	1.71
Decanal	0.73	0.19	-	0.18	0.21
Other aroma	13.32	3.39	1.70	1.99	3.27
Sinensal	7.29	1.50	0.57	0.78	2.08
Rest	24.15	5.47	4.26	6.49	8.46

Table B.6. Extracted compounds during desorption of Spanish red oil at 25 wt.-% loading per sample.

sample	used CO$_2$ (kg)	others (mg)	Citral (mg)	Decanal (mg)	Linalool (mg)	other Mono-terpenes (mg)	Limonene (mg)
1	0.023	2.928	0	0	0	51.040	1657.732
2	0.062	0	0	0	0	32.799	1109.501
3	0.101	0.639	0	0	0	11.615	412.147
4	0.180	0	0	0	0	6.260	244.540
5	0.258	0	0	0	0	7.360	334.540
6	0.336	0.401	0	0	0	5.271	406.328
7	0.414	5.417	1	0	0	0	0.183
8	0.453	17.837	0.343	4.539	1.217	0.079	0
9	0.531	22.158	0	12.121	2.735	0.089	0
10	0.609	23.109	0	6.701	5.187	0.120	0
11	0.687	24.424	0.064	0.929	7.531	0.117	0
12	0.765	13.606	0	0.620	4.892	0.047	0
13	0.843	4.948	0	0.502	1.662	0	0
14	0.921	9.135	0	0.101	4.270	0	0
15	0.999	5.213	0	0.000	2.676	0	0
16	1.189	8.494	0	0.054	3.098	0	0
17	1.640	209.900	0	0	0	0	0

Table B.7. Extracted compounds during desorption of Spanish red oil at 50 wt.-% loading per sample.

sample	used CO$_2$ (kg)	others (mg)	Citral (mg)	Decanal (mg)	Linalool (mg)	other Mono-terpenes (mg)	Limonene (mg)
1	0.014	0	0	0	0	56.993	2125.007
2	0.034	0.000	0	0	0	68.138	2561.262
3	0.068	0.000	0	0	0	56.501	1984.399
4	0.137	2.994	0	0	0	12.446	494.660
5	0.232	7.224	0	0	0	14.429	712.855
6	0.321	0.593	0	0	0	7.138	447.569
7	0.414	2.979	0	0.372	0	2.669	182.479
8	0.444	3.210	0	1.054	0	0	1.036
9	0.485	14.643	0.625	7.106	2.068	0	0.262
10	0.540	14.080	0.544	7.980	2.031	0	0.132
11	0.622	24.844	1.121	11.238	4.247	0	0.156
12	0.738	23.626	1.467	1.758	5.348	0	0.086
13	0.854	22.324	1.604	0.339	6.122	0	0
14	0.977	19.152	1.717	0.197	6.635	0	0
15	1.318	29.897	3.063	0.270	11.645	0	0
16	1.523	18.664	2.189	0.444	4.907	0	0
17	1.728	14.526	0.935	0.149	1.423	0	0
18	2.118	31.485	0.413	0.329	0.889	0	0

Table B.8. Extracted compounds during desorption of Brazilian green oil at 25 wt.-% loading per sample, 1.Run.

No.	CO_2	others	MNMA	Citral	Dec.	Lin.	other mono-terp.	γ-Terp.	Lim.
	(kg)	(mg)	(mg)	(mg)	(mg)	(mg)	(mg)	(mg)	(mg)
1	0.023	0.934	0	0	0	0	2.199	7.771	24.696
2	0.062	1.027	0	0	0	0	4.661	14.327	47.985
3	0.101	4.544	0	0	0	0	14.784	37.576	130.996
4	0.180	3.586	0	0	0	0	17.036	42.278	149.600
5	0.258	28.361	0	0	0	0	124.409	394.784	1343.947
6	0.336	11.149	0	0	0	0	32.331	105.689	381.431
7	0.414	5.168	0	0	0	0	24.229	107.718	377.985
8	0.453	14.244	0	0	0	0	19.730	108.637	356.089
9	0.531	10.448	3.601	0.216	0.651	0.508	0.838	6.410	17.329
10	0.609	17.608	8.109	0.320	1.241	0.620	0	0	0.112
11	0.687	14.768	6.510	0.262	0.645	0.572	0	0	0.211
12	0.765	16.089	7.147	0.302	0.224	0.748	0	0.042	0.133
13	0.843	10.904	5.334	0.217	0.038	0.422	0	0	0.182
14	0.921	9.265	3.764	0.241	0	0.277	0.081	0	0.237
15	0.999	12.226	5.129	0.368	0	0.570	0	0	0.190
16	1.189	19.958	4.630	0.683	0	1.371	0	0	0
17	1.640	20.224	0.739	0.674	0	0.822	0	0	0.281

Table B.9. Extracted compounds during desorption of Brazilian green oil at 25 wt.-% loading per sample, 2.Run.

No.	CO$_2$	others	MNMA	Citral	Dec.	Lin.	other monoterp.	γ-Terp.	Lim.
	(kg)	(mg)	(mg)	(mg)	(mg)	(mg)	(mg)	(mg)	(mg)
1	0.023	1.295	0	0	0	0	2.425	8.373	26.806
2	0.062	1.659	0	0	0	0	8.008	22.227	75.807
3	0.101	35.012	0	0	0	0	151.091	462.529	1569.968
4	0.180	13.606	0	0	0	0	49.247	159.015	561.132
5	0.258	15.379	0	0	0	0	31.530	121.684	426.108
6	0.348	4.335	0.476	0	0	0	11.512	60.309	211.444
7	0.458	3.123	0.696	0	0	0	2.691	18.168	59.919
8	0.553	22.633	0.981	0.286	1.317	0.719	0.536	0.159	0.350
9	0.649	7.689	6.211	0.242	1.042	0.527	0	0	0.144
10	0.724	4.893	4.072	0.161	0.431	0.336	0	0	0.181
11	1.059	11.189	10.994	0.392	0.148	0.440	0	0	0.238
12	1.264	8.786	5.515	0.289	0	0.753	0	0	0.197
13	1.503	9.549	0	0.250	0	0.300	0	0	0
14	2.972	32.274	0	1.089	0	1.837	0	0	0

Table B.10. Extracted compounds during desorption of Brazilian green oil at 50 wt.-% loading per sample.

No.	CO$_2$	others	MNMA	Citral	Dec.	Lin.	other monoterp.	γ-Terp.	Lim.
	(kg)	(mg)	(mg)	(mg)	(mg)	(mg)	(mg)	(mg)	(mg)
1	0.023	35.287	0	0	0	0	93.340	273.594	928.579
2	0.041	73.871	0	0	0	0	292.542	868.855	2957.132
3	0.101	105.23	0	0	0	0	145.258	449.058	1579.658
4	0.180	5.344	0	0	0	0	24.091	92.602	333.063
5	0.258	9.080	0	0	0	0	18.165	91.146	319.009
6	0.336	2.444	0.087	0	0	0	2.992	18.818	58.745
7	0.414	9.600	0.438	0	0	0	0	0	0
8	0.453	8.881	4.738	0.205	1.177	0.558	0.038	0.086	0.123
9	0.531	13.036	10.475	0.460	2.493	1.139	0	0	0
10	0.663	20.655	16.778	0.733	3.206	1.810	0	0.077	0
11	0.717	12.330	12.092	0.512	0.353	1.205	0	0.070	0.131
12	0.820	13.426	13.777	0.616	0	1.457	0	0	0.094
13	0.956	16.051	15.269	0.968	0	1.974	0	0	0
14	1.059	18.330	4.262	0.810	0	1.730	0	0	0
15	1.400	31.883	1.026	1.773	0	3.011	0	0	0.115
16	1.845	32.418	0	1.318	0	1.552	0	0	0.112
17	2.630	41.913	0	1.287	0	0	0	0	0

Appendix C : Experimental Results for Rice Bran Oil

Table C.1. Results obtained for the fractionation of RBO
P=25 MPa, T=60 °C, CO_2-flow =2 kg/h, Feed=0.1289 kg/h.

No.	t (min)	Extract reflux (kg/h)	Extract (kg/h)	Raffinate (kg/h)	FFA (wt.-%)	TG (wt.-%)	Sterols (wt.-%)	Oryzanol (wt.-%)
Extract								
1 E	60	0.68	0.021	0.097	91.29	-	-	-
2 E	90	0.36	0.021	0.097	97.82	-	-	-
3 E	120	0.42	0.021	0.097	75.31	-	-	-
4 E	150	0.57	0.021	0.097	77.40	-	-	-
5 E	180	0.34	0.021	0.097	77.66	n.d.	0.15	0.21
6 E	210	0.44	0.021	0.097	78.53	n.d.	-	-
7 E	240	0.43	0.021	0.097	89.74	-	-	-
Raffinate								
6 R	210	0.44	0.021	0.097	21.54	86.26	0.76	1.7

Table C2. Results obtained for the fractionation of RBO
P=14 MPa, T=60 °C, CO_2-flow =2 kg/h, Feed=0.0612 kg/h.

No.	t (min)	Extract reflux (kg/h)	Extract (kg/h)	Raffinate (kg/h)	FFA (wt.-%)	TG (wt.-%)	Sterols (wt.-%)	Oryzanol (wt.-%)
Extract								
1E	60	0.68	0.007	0.057	94.13	-	-	-
2E	90	0.36	0.007	0.057	85.28	-	-	-
3E	120	0.42	0.007	0.057	90.90	-	-	-
4E	150	0.57	0.007	0.057	76.84	-	-	
5E	180	0.34	0.007	0.057	89.01	1.14	0.23	0.18
6E	210	0.44	0.007	0.057	80.74	-	-	

Table C3. Results obtained for the fractionation of RBO
P=20 MPa, T=60 °C, CO_2-flow =2 kg/h, Feed=0.1289 kg/h.

No.	t (min)	Extract reflux (kg/h)	Extract (kg/h)	Raffinate (kg/h)	FFA (wt.-%)	TG (wt.-%)	Sterols (wt.-%)	Oryzanol (wt.-%)
Extract								
1E	60	0.68	0.024	0.103	100.00	-	-	-
2E	90	0.36	0.024	0.103	92.88	-	-	-
3E	120	0.42	0.024	0.103	81.55	-	-	-
4E	150	0.57	0.024	0.103	92.04	-	-	-
5E	180	0.34	0.024	0.103	87.05	-	-	-
6E	210	0.44	0.024	0.103	84.64	0	0.26	0.19

Table C4. Results obtained for the fractionation of RBO
P=14 MPa, T=80 °C, CO_2-flow =2 kg/h, Feed=0.1289 kg/h.

No.	t (min)	Extract reflux (kg/h)	Extract (kg/h)	Raffinate (kg/h)	FFA (wt.-%)	TG (wt.-%)	Sterols (wt.-%)	Oryzanol (wt.-%)
Extract								
7E	60	0.68	0.028	0.101	82.19	-	-	-
8E	90	0.36	0.028	0.101	75.70	-	-	-
9E	120	0.42	0.028	0.101	81.06	n.d.	0	-
10E	150	0.57	0.028	0.101	-	-	-	-
11E	180	0.34	0.028	0.101	-	-	-	-
12E	210	0.44	0.028	0.101	75.36	n.d.	0	0.17

Table C5. Results obtained for the fractionation of RBO
P=20 MPa, T=80 °C, CO_2-flow =2 kg/h, Feed=0.1289 kg/h.

No.	t (min)	Extract reflux (kg/h)	Extract (kg/h)	Raffinate (kg/h)	FFA (wt.-%)	TG (wt.-%)	Sterols (wt.-%)	Oryzanol (wt.-%)
Extract								
13E	60	0.68	0.024	0.103	100.6	-	-	-
14E	90	0.36	0.024	0.103	100.5	-	-	-
15E	120	0.42	0.024	0.103	96.68	-	-	-
16E	150	0.57	0.024	0.103	89.47	-	0	-
17E	180	0.34	0.024	0.103	82.61	0.72	0	0.25
18E	210	0.44	0.024	0.103	89.67	-	-	-

Table C6. Results obtained for the fractionation of RBO
P=25 MPa, T=80 °C, CO_2-flow =2 kg/h, Feed=0.1289 kg/h.

No.	t (min)	Extract reflux (kg/h)	Extract (kg/h)	Raffinate (kg/h)	FFA (wt.-%)	TG (wt.-%)	Sterols (wt.-%)	Oryzanol (wt.-%)
Extract								
19E	60	0.68	0.027	0.099	76.40	-	-	-
20E	90	0.36	0.027	0.099	98.53	-	-	-
21E	120	0.42	0.024	0.103	108.9	-	-	-
22E	150	0.57	0.027	0.099	104.7	-	-	-
23E	180	0.34	0.027	0.099	77.45	-	0	-
24E	210	0.44	0.027	0.099	94.21	0.56	0	0.19

Table C7. Results obtained for the fractionation of RBO
P=17 MPa, T=80 °C, CO_2-flow =2 kg/h, Feed=0.1289 kg/h.

No.	t (min)	Extract reflux (kg/h)	Extract (kg/h)	Raffinate (kg/h)	FFA (wt.-%)	TG (wt.-%)	Sterols (wt.-%)	Oryzanol (wt.-%)
Extract								
25E	60	0.68	0.026	0.098	77.21	-	-	-
26E	90	0.36	0.026	0.098	88.21	-	-	-
27E	120	0.42	0.026	0.098	79.27	-	-	-
28E	150	0.57	0.026	0.098	81.01	-	0	0.2
29E	180	0.34	0.026	0.098	87.35	0.3	0	-
30E	210	0.44	0.026	0.098	77.86	-	-	-

Table C8. Results obtained for the fractionation of RBO
P=22 MPa, T=60 °C, CO_2-flow =2 kg/h, Feed=0.1289 kg/h.

No.	t (min)	Extract reflux (kg/h)	Extract (kg/h)	Raffinate (kg/h)	FFA (wt.-%)	TG (wt.-%)	Sterols (wt.-%)	Oryzanol (wt.-%)
Extract								
31E	60	0.68	0.024	0.103	88.04	-	-	-
32E	90	0.36	0.024	0.103	91.31	-	-	-
33E	120	0.42	0.024	0.103	96.26	-	-	-
34E	150	0.57	0.024	0.103	91.50	0.69	-	0.24
35E	180	0.34	0.024	0.103	87.53	-	-	-
36E	210	0.44	0.024	0.103	96.54	-	-	-